TOWARD SUSTAINABILITY

Soil and Water Research
Priorities for
Developing Countries

Committee on International Soil and Water
Research and Development

Water Science and Technology Board
Commission on Engineering and Technical Systems

Board on Science and Technology for International Development
Office of International Affairs

National Research Council

NATIONAL ACADEMY PRESS
Washington, D.C. 1991

NOTICE: The project that is the subject of this report was approved by the Governing Board of the National Research Council, whose members are drawn from the councils of the National Academy of Sciences, the National Academy of Engineering, and the Institute of Medicine. The members of the committee responsible for the report were chosen for their special competencies and with regard for appropriate balance.

This report has been reviewed by a group other than the authors according to procedures approved by a Report Review Committee consisting of members of the National Academy of Sciences, the National Academy of Engineering, and the Institute of Medicine.

The National Academy of Sciences is a private, nonprofit, self-perpetuating society of distinguished scholars engaged in scientific and engineering research, dedicated to the furtherance of science and technology and to their use for the general welfare. Upon the authority of the charter granted to it by the Congress in 1863, the Academy has a mandate that requires it to advise the federal government on scientific and technical matters. Dr. Frank Press is president of the National Academy of Sciences.

The National Academy of Engineering was established in 1964, under the charter of the National Academy of Sciences, as a parallel organization of outstanding engineers. It is autonomous in its administration and in the selection of its members, sharing with the National Academy of Sciences the responsibility for advising the federal government. The National Academy of Engineering also sponsors engineering programs aimed at meeting national needs, encourages education and research, and recognized the superior achievements of engineers. Dr. Robert M. White is president of the National Academy of Engineering.

The Institute of Medicine was established in 1970 by the National Academy of Sciences to secure the services of eminent members of appropriate professions in the examination of policy matters pertaining to the health of the public. The Institute acts under the responsibility given to the National Academy of Sciences by its congressional charter to be an adviser to the federal government and, upon its own initiative, to identify issues of medical care, research, and education. Dr. Stuart Bondurant is acting president of the Institute of Medicine.

The National Research Council was organized by the National Academy of Sciences in 1916 to associate the broad community of science and technology with the Academy's purposes of furthering knowledge and advising the federal government. Functioning in accordance with general policies determined by the Academy, the Council has become the principal operating agency of both the National Academy of Sciences and the National Academy of Engineering in providing services to the government, the public, and the scientific and engineering communities. The Council is administered jointly by both Academies and the Institute of Medicine. Dr. Frank Press and Dr. Robert M. White are chairman and vice chairman, respectively, of the National Research Council.

Support of this project was provided by the Agency for International Development No. DPE-5545-A-00-8068-00. The U.S. Agency for International Development reserves a royalty-free and nonexclusive and irrevocable right to reproduce, publish, or otherwise use and to authorize others to use the work for government purposes.

Library of Congress Catalog Card No. 91-67071
ISBN 0-309-04641-6

A limited number of copies are available from:

The Water Science and Technology Board
National Research Council
2101 Constitution Avenue, NW
Washington, D.C. 20418

Board on Science and Technology for
 International Development
National Research Council
2101 Constitution Avenue, NW
Washington, D.C. 20418

Additional copies are available for sale from:

National Academy Press
2101 Constitution Avenue, NW
Washington, D.C. 20418

S–483

Printed in the United States of America

COMMITTEE ON INTERNATIONAL SOIL AND WATER RESEARCH AND DEVELOPMENT

LEONARD BERRY, *Chair*, Florida Atlantic University, Boca Raton
SUSANNA B. HECHT, University of California, Los Angeles
CHARLES W. HOWE, University of Colorado, Boulder
JACK KELLER, Utah State University, Logan
CHARLES McCANTS, North Carolina State University, Raleigh
HUGH POPENOE, University of Florida, Gainesville
PAUL TENG, International Rice Research Institute, Manila, Philippines
    (through September 1990)
GORO UEHARA, University of Hawaii, Honolulu

Staff

CHRIS ELFRING, *Project Director*
PATRICIA CICERO, *Project Assistant*

JACQUELINE MACDONALD, *Research Associate*
JEANNE AQUILINO, *Administrative Specialist*
ANITA A. HALL, *Administrative Secretary*
PATRICIA CICERO, *Project Assistant*

Contents

TOWARD SUSTAINABILITY

Soil and Water Research

Priorities for

Developing Countries

Summary

One of the challenges we face in developing agricultural strategies that are truly sustainable is maintaining the resource base—the soil and water that make agriculture possible. But the pressures on these resources are extraordinary: five billion people now inhabit the earth, with an additional one billion expected each decade well into the next century. The specter of possible changes in climate adds another level of uncertainty. It is time to ask how we can move "toward sustainability," toward a strategy of agriculture and natural resource management that supports current populations while leaving future generations an equitable share of the earth's great wealth.

Population growth, intensified land use, environmental degradation, and agricultural productivity are interrelated issues. Although agricultural technology has performed well in the last 20 years to meet the needs of a vastly larger and generally more prosperous world population, there is concern that those initiatives have peaked and that the technologies in use focus mainly on the best sites—flat areas with ample water and few soil constraints. Scientists in the United States and throughout the world are worried about the decline in productivity in many soil and water systems, especially in the high population growth regions—Africa, Asia, and Latin America.

Meeting the world's increased needs and expectations will require concerted effort. Research is necessary on three fronts. First, techniques must be developed to intensify use of good quality lands while minimizing environmental degradation. Second, ways must be sought to enhance production and reduce degradation on lands previously viewed as "marginal" or "ecologically fragile." Finally, new emphasis must focus on restoring degraded lands.

It is within this context that the Agency for International Development (AID) asked the National Research Council (NRC) "to develop a broad agenda for directing worldwide international research and development efforts related to the use of soil and water resources to sustain agriculture." The NRC's Committee on Soil and Water Research and Development (CISWRD) was established in 1990 to prepare this report.

While we have gained a much better understanding of soil and water systems in the developing world over the past few decades, too little of this new knowledge has been successfully applied to many fundamental management problems. There are substantial gaps in our basic understanding of the ecology of these systems and of the social complexity inherent in resource use.

The committee began its work with a clear charge to look at soil and water research priorities that would contribute to sustainable agricultural strategies. Sustainable agriculture had become a major policy thrust for AID and many other organizations in the development community. But it soon became clear that the rift that often separates those interested in agriculture from those interested in natural resources needs now—more than ever—to be bridged. If sustainability is a goal, then agriculture and natural resource management interests must recognize that they are equal partners in the effort. Competition for "ownership" of the issue is counter productive. These interests must be willing to negotiate a coordinated strategy that includes the strengths of both orientations.

The most compelling theme that emerged during this study is the need for better integration of soil and water research with other elements relevant to natural resource management. Soil and water practices are not independent endeavors, but rather must be an integral part of a larger landscape management. Our understanding of the basic principles of soil and water processes is fairly good, but our ability to apply this knowledge to solve problems in complex local and cultural settings is weak. The single issue research approaches of the past brought great benefits, but the problems we face are changing and demand a more holistic vision.

A FRAMEWORK FOR ASSESSING RESEARCH PRIORITIES

To achieve sustainable agriculture, the world's agricultural productivity must be enhanced while its resource base is conserved. Research will be essential to this task. The complex nature of sustainability demands that the research entail a systems approach that includes integrated research design, interdisciplinary and farmer participation, and a broad perspective as well as specific focuses. A systems-based framework needs to be devised so future research—whether guided by the priorities outlined in this report or elsewhere—can be effective and efficient.

A first step toward sustainability is the matching of organisms and husbandry to the in-site characteristics of the land and water environment and, of course, to the resource preferences and economic and cultural context of the users. This is an obvious, but often difficult, task. The aim of any research framework should be to identify the important elements of these mismatches—the most significant constraints on sustainability—and resolve them.

CRITICAL RESEARCH PRIORITIES

Two key indicators of deterioration in agricultural systems are declines in the quality of the soil and the water. Poor management of either of these resources quickly leads to decreases in farm productivity. Most developing countries occupy tropical zones ranging from seasonally arid to humid tropical environments. Agriculture in tropical environments faces different constraints than in temperate regions, and this affects soil and water research needs.

Given the problems faced by tropical agriculture, the unique characteristics of the environments and cultures, and the strengths and weaknesses of the existing data base, research in the following six areas could offer great rewards in support of sustainable agriculture and natural resource management:

- Overcoming institutional constraints on resource conservation;
- Enhancing soil biological processes;
- Managing soil properties;
- Improving water resource management;
- Matching crops to environments; and
- Effectively incorporating social and cultural dimensions into research.

To further these goals, the wealth of time-tested indigenous knowledge that exists needs to be tapped. Special potential lies in the blending of traditional and modern knowledge. One of the most intractable problems yet to be faced is the difficulty of communicating new ideas to the farmer and establishing two-way communication between farmers and researchers. Research and development organizations have struggled with this problem for many years, and it remains a high priority issue.

AN INTEGRATED RESEARCH STRATEGY

A collaborative, integrated research strategy requires institutional mechanisms and structures that effectively link research efforts and organizations with clients, and that enhance the interactions among the different components of research. Mechanisms are needed to reassess research priorities periodically and to generate local data about soil and water resources.

A basic issue in any attempt to target research to the needs of users is the pattern of communication and feedback among the different people involved.

The chains of communication can be complex. Traditionally, crop research went through a hierarchial sequence from basic research to field testing to extension-agent promotion. But this structure has not always worked in developing countries. Special efforts are required to encourage networks, "intermediate change agents" (e.g., private voluntary organizations), and other mechanisms to link researchers and research organizations with universities in host countries, private voluntary organizations, village organizations, and farmers in interactive exchanges. Participation from the ultimate recipients of research—the farmers—is needed throughout the process of planning and conducting research.

CONCLUSIONS

Some common themes crystallized during the committee's deliberations:

• Major gaps still exist in our understanding of soil and water systems and processes, but more important are the gaps between what is known and what is applied.

• Indigenous knowledge should always be assessed. It often can suggest promising research on ecosystem components and strategies, such as nitrogen fixing trees, nutrient accumulating species, and low input irrigation techniques. In some cases, it can provide a platform for the integration of traditional and new technologies.

• More effective links between the social and the natural science aspects of soil and water problems are needed. Social and economic contexts create constraints that can effectively limit the application of technical improvements unless such contexts are adequately understood and addressed.

• More effective ways to use research resources for long-term, practical ends are needed. How can better feedback and communication be established between the field and the research institution so research can be focused on real, practical problems?

• The weakest link in the research process is the dissemination of research findings to the farm or regional levels, with the great physical and human diversity that occurs. Greater effort is needed to develop better ways to communicate results.

Soil and water resources provide the foundation upon which agriculture is based. But successful agricultural production systems require a combination of biological and societal resources. This is a complex and dynamic mix of variables. In view of the evolutionary nature of agricultural systems, it is important that the setting of research priorities be an ongoing process. Research priorities must be reassessed and adjusted periodically to serve the problems at hand. A mechanism is needed for evaluating and reiterating priorities to keep them fresh, flexible, and responsive to current needs.

An effective effort to build sustainability into our agricultural systems will require changes in the philosophy and operating procedures of development organizations. Program planners and implementers will need to be more responsive to the evolution of individual agricultural systems and to the broader aspirations, needs, and capabilities of the user populations.

The search for ways to achieve sustainable agriculture and natural resource management will require changes in our traditional approach to problem solving. Researchers must cross the boundaries of their individual disciplines; they must broaden their perspective to see the merits of indigenous knowledge; and they must look to the farmer for help in defining a practical context for research. This change in vision is under way in various degrees throughout the research community, but the pace of change is slow.

1

Introduction

> Soil and water are critical components of the resource base upon which agriculture depends. To move toward sustainability, agriculture and natural resource management interests must recognize that they are equal partners in the effort. The challenge is to adapt and extend our knowledge about soil and water to develop economically productive, culturally appropriate, and environmentally sound agricultural systems. A flexible, ongoing process is necessary to set research priorities to support inherently dynamic agricultural systems.

In this final decade of the 20th century, there is a penchant for looking forward to a new millennium. The turning of the calendar in such a significant way offers a natural pause—an opportunity to reflect on the past and to anticipate the future. This report looks to the future with emphasis on the soil and water resources that support agriculture. With world population now more than five billion, and some three billion people entering their reproductive years, it is a timely moment for such an assessment. Population growth and economic demands are exerting mounting pressures on the earth's soil, water, and other natural systems. The specter of possible changes in climate can only add to these pressures. It is time to consider how we can move "toward sustainability," toward a vision of natural resource management that supports current populations while leaving future generations an equitable share of the earth's great wealth.

Although the most alarmist headlines portending environmental doom are probably overdrawn, many scientists are concerned about the steady decline in basic productivity of many soil and water systems, particularly those in the high population growth regions—Africa, Asia, and Latin America. In the humid tropics, rain forests are converted to agriculture and, in some places, large-scale ranching, bringing losses of topsoil and the depletion of nutrients, especially nitrogen (Lal, 1988; Pimentel et al., 1987; Sanchez and

Benites, 1987). In arid and semiarid areas, increased demands for food, fodder, fuel, and materials bring analogous degradation (NRC, 1984). In hill lands, problems are exaggerated because the slopes accentuate runoff and erosion (Jodha, 1990).

Soil compaction and crusting, loss of soil organic matter, reduced activity by soil organisms, nutrient deficiencies and imbalances—these are examples of the many forms of soil resource degradation. The interrelated issues of population growth, economic pressures, intensified land use, and environmental degradation at local and regional levels are serious causes for concern. Although these issues are universal, they are of particular concern in the developing nations of the tropics where the economic constraints of largely agrarian societies and the needs of expanding human populations are most pressing (NRC, 1991b). Some experts worry that continued population growth may make sustainable development all but impossible (Doyle, 1991).

While current agricultural and biological technology has performed well in the last 20 years in meeting the growing needs of a vastly larger world population, there is concern that those initiatives are losing energy. In

Degradation of soil and water resources is a serious concern throughout the world. In the Philippines, as in mountainous regions everywhere, deforestation of the uplands leads to soil loss and sedimentation of waterways. *Credit*: Michael McD. Dow, National Research Council.

addition, pressures are mounting on more environmentally fragile lands, and this trend will surely continue.

A three-pronged approach will be necessary to increase agricultural production, with research supporting each path. First, farmers must intensify their use of good quality lands, with an eye toward ameliorating problems in areas with the best resources. Second, farmers will be forced to expand, maintain, and enhance production on lands previously viewed as "marginal"—such as steep lands, tropical forest zones, and the semiarid tropics. When misused, such lands often experience high levels of degradation, yet they already support a significant portion of the world's population. Finally, new emphasis will need to be focused on restoring degraded lands.

It will truly be a challenge to increase or intensify production on environmentally fragile lands—lands that traditionally have been judged economically unsuitable for agriculture. Without significant industrial input, these lands often have low productivity potential and, in any case, are highly vulnerable to degradation; many areas already require some form of recuperative intervention. Environmental concerns in these areas are of special importance where they encompass high biological and cultural diversity, including some areas of extraordinary endemic resources. These areas demand a delicate approach to both research and technical interventions. They require a research thrust that focuses on sustainability, both in terms of economics and natural resource management. Local participation in the design of research goals is essential.

It is within the context of these realities, therefore, that the Agency for International Development (AID) asked the National Research Council "to develop a broad agenda for directing worldwide international research and development efforts related to the use of soil and water resources to sustain agriculture, outlining both short- and long-term priorities." The NRC's Committee on Soil and Water Research and Development (CISWRD) was established in 1990 to undertake this task and prepare a report.

Such a study is timely because while we have gained a much better understanding of soil and water systems in the developing world over the past few decades, too little of this new knowledge has been successfully applied to fundamental management problems. There also are gaps in our basic understanding of the ecology of these systems and of the social complexity inherent in resource use.

The gravity of these issues is apparent from the attention being focused on sustainable agriculture and sustainable natural resource management. Even within the National Research Council, two committees in addition to CISWRD are active in these areas. The Committee on Sustainable Agriculture and the Humid Tropics is charged to identify and analyze agricultural practices that contribute to environmental degradation and declining agricultural production in humid tropical environments worldwide. The Committee for Collaborative Research Support for AID's Sustainable Agriculture and Natural Resource Management Program (SANREM) focused on the task of designing a Collaborative Research Support Program dedicated to agriculture and natural resource management.

This committee began its work with a clear charge to look at soil and water research priorities that would contribute to "sustainable agriculture." The quest for sustainable agriculture had become a major policy thrust for the Agency for International Development and many other organizations in the development community. But it was soon clear to the committee that the rift that often separates those interested in agriculture from those interested in natural resources needs now—more than ever—to be bridged. For if sustainability is a goal, agriculture and natural resource management interests must recognize that they are equal partners in the effort. Competition for "ownership" of the issue is counter productive. These interests must be willing to negotiate a coordinated strategy that includes the strengths of both orientations.

Rather than debate the definition of sustainable agriculture,[1] the committee elected to accept the definition in use at AID while our work was occurring (Department of State, 1990): "sustainable agriculture is a management system for renewable natural resources that provides food, income, and livelihood for present and future generations and that maintains or improves the economic productivity and ecosystem services of these resources."

This definition carries several implications. It requires sustainable agricultural systems to be both economically and ecologically viable. Management choices must give priority attention to maintaining the renewable resource base and its ability to meet the changing needs of humankind. The definition also recognizes that natural resources perform ecosystem services beyond the production of food, fiber, fuel, and income. These additional contributions include the recycling of nutrients, detoxification of noxious chemicals, continuation of evolutionary processes, purification of water, and regulation of the hydrological processes within watersheds and across the landscape.

---

[1]The definition of agricultural sustainability, it is frequently noted, varies by individual, discipline, profession, and area of concern. The literature offers hundreds of definitions of sustainable agriculture. Virtually all definitions, however, incorporate the following characteristics: long-term maintenance of natural resources and agricultural productivity, minimal environmental impacts, adequate economic returns to farmers, optimal production with minimized chemical inputs, satisfaction of human needs for food and income, and provision for the social needs of farm families and communities. All definitions, in other words, explicitly promote environmental, economic, and social goals in their efforts to clarify and interpret the meaning of sustainability. In addition, all definitions implicitly suggest the need to ensure flexibility within the agroecosystem in order to respond effectively to stresses (NRC, 1991b).

The committee originally consisted of eight scientists, and despite a diversity of backgrounds its members soon realized their limitations in terms of topical and regional coverage (see Appendix A). Thus we sought ways to bring more diverse expertise into the process. The committee elected to host a workshop to solicit additional ideas, information, and strategies. The two-day workshop was held October 1-2, 1990, at the National Academies' Beckman Center in Irvine, California. The goal was to bring together a range of professionals in critical aspects of soil and water management, and use their input to develop a research agenda and priorities to help AID and other international development organizations plan an efficient strategy for promoting the use of soil and water to sustain agriculture.

Approximately 30 scientists representing a wide range of disciplines in the physical, biological, and social sciences participated (see Appendix B). The workshop was structured to focus on the elements essential to sound soil and water management in the context of systems dynamics. Four working groups were organized to give attention to the following topics:

- The biotic environment;
- Water resources;
- Physical properties of soils; and
- Chemical properties of soils (salinity, acidity, nutrients).

In the past, research on soil and water resources tended to focus on components (e.g., soil physics in irrigated systems) while neglecting the linkages among biotic, physical, and human factors that shape the way the agricultural system functions and determine how it adapts to stress. Committee members and workshop participants alike, however, stressed the need for a more integrative, systems approach both in this report and in soil and water research. Thus, even though the workshop structure at times divided participants into focused groups, the emphasis was on interactions across disciplines and ecosystems.

Each group was chaired by a member of CISWRD and included participants with expertise in indigenous knowledge and agroecology, and specialists in the dissemination of agricultural and scientific information. The working groups were given four tasks:

1. Discuss the nature of the assigned subset of the soil and water arena, including problems, scope, components, and relative importance;

2. Review the state of the art of the related knowledge base, the adequacy of the knowledge base, and the degree to which the knowledge base is actually being used;

3. Reflect on the priorities for research and development that are most appropriate within the topic; and

4. Discuss communications as it affects, or could affect, the topic, including information synthesis and dissemination, possible networks, and alternative communications procedures.

Several plenary sessions were included to facilitate close coordination among groups and to seek links between the various topics. Vital contributions were made by AID staff, including William Furtick, David Bathrick, James Bonner, and Thurman Grove. The workshop provided the committee with valuable resources for its subsequent attempt to outline an effective framework and suggest priorities for soil and water research.

A recurring point that emerged during the committee's tenure was the need to integrate soil and water research with other elements of natural resource management. Soil and water practices are not stand-alone endeavors but rather are integral components of a total management scheme. Our understanding of the basic principles is fairly strong, but our ability to apply this knowledge to solve problems in complex local and cultural settings is weak. Experimental approaches that do not involve more systematic, integrated research strategies will not be particularly useful for supporting long-term sustainability. The reductionist approaches of the past brought great benefits, but the problems we face are changing and demand a more holistic vision.

Some common themes emerged during the committee's deliberations, and these will be discussed in more detail in subsequent chapters:

• Major gaps still exist in our understanding of soil and water systems and processes, but more important are the gaps between what is known and what is applied.

• Indigenous knowledge and practice should always be assessed. It often can suggest promising research on ecosystem components and strategies, such as nitrogen fixing trees, nutrient accumulating species, and low input irrigation techniques. In some cases, it can provide a platform for the integration of traditional and new technologies.

• More effective links between the social and the natural science aspects of soil and water problems are needed. Social and economic contexts create constraints that can effectively limit the application of technical improvements unless such contexts are adequately understood and addressed.

• More effective ways to use research resources for long-term, practical ends are needed. How can better feedback and communication be established between the field and the research institution so research can be focused on real, practical problems?

• The weakest link in the research process is the dissemination of research findings to the farm or regional levels, given the great local physical and cultural diversity that occurs. Greater effort is needed to develop better ways to communicate results.

Soil and water resources provide the foundation upon which agriculture is based. But it is a combination of biological and societal resources that are required to make successful agricultural production systems. This is a complex and dynamic mix of variables. In view of the evolutionary nature of agricultural systems and our knowledge of them, it is important that the setting of research priorities be an ongoing process. Research priorities must be reassessed and adjusted periodically to serve the problems at hand. Thus, a mechanism is needed for evaluating and establishing priorities to keep them fresh, flexible, and responsive to current needs.

The challenge is to adapt and extend our scientific knowledge about soil and water to establish economically productive, environmentally sustainable agricultural systems. Sustainability requires careful selection of the production systems and development of associated husbandry programs that are sustainable within, and do not destroy, the resource base.

An effective effort to build sustainability into our agricultural systems will require changes in the philosophy and operating procedures of development organizations. Program planners and implementers will need to be more responsive to the evolution of individual agricultural systems and to the broader aspirations, needs, and capabilities of the target populations. We believe this report can be useful to the scientific and development community during this transition and hope that it will stimulate comment and debate.

2

A Framework for Assessing Research Priorities

Soil and water research must focus on issues that will enable farmers to manage natural resources to gain maximum efficiency of use while maintaining or enhancing environmental quality. A common constraint is the "mismatch" between the characteristics of the soil and water environment and the desired use of the site; the challenge is to identify and employ appropriate management practices. The range of options needed can best be developed using a systems approach, which can facilitate relatively precise evaluations of problems, solutions, and results.

It is not surprising that a small committee specifically charged to examine the complex problem of setting soil and water research priorities should decide that new approaches, as well as new priorities, are needed. It is remarkable, however, that a gathering of 30 scientists chosen primarily for their expertise on individual parts of the soil and water research agenda should agree, almost without exception, that the most critical priority is not any one area but rather the links among areas. The primary reason for this departure from tradition could be that the guiding question was not just "what are the top priority soil and water research issues" but rather, "how can soil and water research play a significant role in managing agroecosystems for sustainability"—a very different task.

Achieving sustainable agriculture will demand that the world's agricultural production capacity be enhanced while its resource base is conserved. If the well-being of the world's less advantaged people is to improve in any lasting sense, long-range concerns about security and the health of natural resources must be addressed when planning future economic and social development. Research will be essential to this task. Researchers must devote greater attention to developing integrated cropping, livestock, and other production systems—and the specific farming practices within these systems—that enhance (or, at minimum, do not degrade) the structure and functioning of the broader agroecosystem and regional landscapes (NRC, 1991b).

A primary objective of research on sustainable agriculture and natural resource management is the integration of information in its application to the problems of agricultural development (Edwards et al., 1990; Grove et al., 1990). This process requires an approach to interdisciplinary research that includes the following: (1) identification of the components and interactions that determine the structure and functioning of the agroecosystem as a whole; (2) formulation of hypotheses that focus on those components and interactions within the entire agroecosystem; (3) examination, testing, and measurement of the hypotheses; and (4) interpretation of results as they pertain to the various components of the agroecosystem and to the system as a whole. A lack of understanding of the interrelatedness of system components has undermined agricultural sustainability in the past (NRC, 1991b).

In the United States, the dearth of systems research and political will has been identified as a key obstacle to the adoption of alternative farming practices and as necessary to the development of more sustainable agriculture (NRC, 1989a, 1989b). The integrated research design, interdisciplinary participation, and systemwide perspective that the systems approach entails are even more necessary elsewhere in the world if the complex nature of sustainability is to be understood and threats to sustainability identified and addressed.

A systems-based framework needs to be devised so that future research—whether guided by the priorities outlined in this report or elsewhere—can be effective, efficient, and focused. Framework here means a structure of ideas, a guiding vision, under which research priorities are set—the goals, objectives, and program mission of the underlying organizational structures. It is the framework, rather than any particular set of priorities, that will have lasting impact as long as the system has the capacity to receive feedback from the field and translate it into guidance for action.

This report is an overview and thus general where some might hope for concrete. It is not intended to be a simple road map because that is not possible, given the issue. This chapter sets out the committee's thinking on an appropriate framework for setting soil and water research priorities. The central problem is to integrate a spectrum of component-oriented research results and focus the knowledge on the problem of sustainability, particularly on marginal lands. At the same time, there is a need to bring local wisdom together with modern scientific knowledge and require that at least some portion of future research be driven by problems identified in the field. This suggests the need for effective feedback mechanisms that include farmer participation throughout the design and implementation of a research strategy.

KEY ELEMENTS IN THE RESEARCH FRAMEWORK

Sustainability of production at a given location is obtained by appropriate management of all elements of that environment at that site. In some cases, the activities will be conservation oriented, while in others production might be the focus. In any case, the management strategy must exist in the context of the local landscape, cultural concerns, and land use practices. To ensure adequate and increasing productivity over the long term, the use of water, soil, and other natural resources needs to be understood within the evolving social and economic context.

A first step toward sustainability is the matching of organisms and husbandry to the in-site characteristics of the land and water environment and, of course, to the resource preferences and characteristics of the users. This is an obvious, but difficult, task and many times "mismatches" occur. Crops are planted that are not suited to the existing soil and water conditions; varieties are used that fail to produce consistently enough to satisfy the

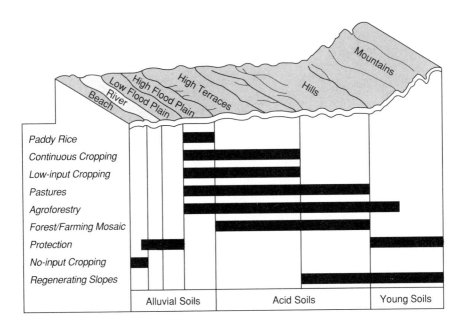

Matching crops to environments requires careful attention to the physical setting, as illustrated in this example of the agricultural practices available for a variety of humid tropical landscapes. In addition, efforts must be made to match crops and production strategies to social, cultural, and economic environments as well. *Credit:* Pedro Sanchez and Jose Benites, North Carolina State University.

needs of the community; increased production pressures leave even the best of local management techniques incapable of replacing nutrients at a rate that ensures sustainability. For these and many other reasons, mismatches occur between the results of current practice and the goal of long-term sustainability. The aim of any research framework is to identify the important elements of these mismatches—the most significant constraints on sustainability—and resolve them, where possible, through the application of current knowledge, appropriate policy, and political action, and where not possible now, through the design of appropriate research. A flexible research framework will do the following:

• Encourage continued research on the specific elements and characteristics of soil and water systems, but in the context of the integrated priorities;
• Take the results of past and present component research and integrate them to focus on current problems;
• Incorporate both scientific and local knowledge into this integration;
• Develop mechanisms to respond to problems identified in the field; and
• Encourage dissemination of research results to the field and monitor the impacts to provide feedback to researchers.

SELECTING APPROPRIATE ORGANISMS AND HUSBANDRY

Throughout the history of agriculture, there have been only three methods used to select appropriate techniques for agricultural systems: trial and error, analogy, and systems analysis. Trial and error has been, and still is, the most common process. People have experimented with agroecosystems over long periods of time and adapted their management to local contexts. Although no local system is perfect, in the past such ongoing "folk" experimentation has served people well. The problem is that changes in population, markets, and tenure systems, and other modern pressures often destabilize these systems. Change occurs at a faster rate than can be accommodated through gradual trial and error. Indigenous knowledge thus is not always appropriate to the environment of use, nor effective in offering the full range of alternatives needed.

Analogy—the selection of potentially useful crops or animals for particular locales by comparing what had been successful at other, similar locales—has been a dominant force in the introduction of crops, animals, and practices from one part of the world to another. In this century, the most prominent example of this process is the spread of high-yielding varieties of rice. Specialized organisms, such as vetiver grass, are being tested on this basis. Analogy will continue to be used as a way to detect the kind of local experimental testing that seems most effective.

Systems analysis—for example, the prediction of the performance of a

TABLE 2.1 An Example of a Minimum Physical Data Set to Predict Crop Performance

Daily weather data:	Soil data by horizons:
• maximum temperature	• sand, silt, and clay content
• minimum temperature	• bulk density
• precipitation	• organic carbon content
• solar radiation	• nitrogen content
	• pH
Crop data:	
• genetic coefficients for development	
• genetic coefficients for growth	
	Management opportunities:
Site information:	• row spacing
• temperature	• plant population
• solar radiation	• planting date
• day length	• irrigation
• soil water deficits	• fertilizer applications
• soil nitrogen deficits	• residue management

Source: International Benchmark Sites Network, 1988.

specific cultivar based on a comparative analysis of the environmental conditions—is the next step in progressively more analytical attempts to match agricultural techniques to the local environment (see Table 2.1). It can provide a more global, process-oriented approach to identifying factors that are key to the sustainability of agroecosystems. Systems analysis has the potential to cut across conventional ecological boundaries and carry knowledge from one agroecosystem to another. It has the capacity to allow quantitative projections of the results of research and to bring a wide range of different kinds of knowledge to bear on the particular problems at hand. This approach can be expanded to include economic, institutional, and cultural, as well as physical environmental site conditions.

APPLYING SYSTEMS ANALYSIS TO
THE RESEARCH FRAMEWORK

To propose systems analysis to a study of agroecosystems is hardly new. What is new, however, is the attempt to bring a level of precision to this integration that would allow relatively precise evaluations of problems, solutions, and results. In attempting to use a systems approach, three major

SYSTEM PROPERTIES

Agroecosystems demonstrate certain system properties that can help researchers understand the complexity inherent in sustainable agricultural systems. These properties can be labeled productivity, stability, sustainability, equitability, diversity, and adaptability (Cuc et al., 1990; KEPAS, 1985; Marten and Rambo, 1988). Although the terminology developed to name and describe these properties may not satisfy everyone, the concepts are useful. These properties are essentially descriptive; they can be used to summarize the behavior of agroecosystems and as indicators of performance.

Productivity is the system's net output of goods and services, commonly measured either as annual yield or net income per unit of input or resource (for example, yield per hectare).

Stability is a measure of the constancy of the productivity. It describes the degree to which productivity remains constant despite normal, small-scale fluctuations in environmental variables (for example, climate or economic conditions). A small degree of variability indicates a high level of stability; a high degree of variability indicates a low level of stability.

Sustainability is the ability of a system to maintain productivity when subjected to stress and shock. A stress is a regular, sometimes continuous and cumulative, relatively small and predictable disturbance (for example, the effect of erosion). A shock, by contrast, is an irregular, infrequent, relatively large, and unpredictable disturbance, such as might be caused by a flood, new pest, or political upheaval. A highly sustainable system is able to recover rapidly and completely from disturbances; a moderately sustainable system can recover, but slowly. A system with low sustainability might collapse or recover only to a lower level of productivity.

Equitability is a term used to express how evenly or fairly the products of an agroecosystem (food, fiber, fuel, income) are distributed among its human beneficiaries at the household, farm, village, regional, or national levels.

Diversity is a measure of the number of different types of components (for example, species) within a system. Diversity allows rural people to spread risks and maintain a minimum level of subsistence even when some activities fail.

Adaptability refers to the ability of the system to respond to change in its environment. This concept is related to the concepts of stability and sustainability. It describes the capacity of an agroecosystem to respond to perturbations and still function at an acceptable level of productivity.

Agricultural development almost inevitably involves trade-offs between the different system properties. For instance, the introduction of a new technology such as fertilizer may have the immediate effect of increasing productivity, but this is often at the expense of one or more other properties. A sense of the dynamic interactions of these properties as agricultural systems are pressured, and how far they can be pushed before the overall sustainability is compromised, is a central question.

issues surface: our understanding of processes, the adequacy of the data base, and the translation of global knowledge to site-specific situations.

A simple conceptual model for the conduct of integrated agricultural systems research would include the following elements (NRC, 1991b):

• Description of the target agroecosystem, including its boundaries and components, functions, interactions among its components, and interactions across its boundaries;
• Analysis of the agroecosystem to determine constraints on, and factors that can contribute to, the attainment of social, economic, and environmental goals;
• Identification of interventions and actions to overcome the constraints;
• On-farm experimentation with interventions; and
• Evaluation of the effectiveness of newly designed systems, and redesign as necessary.

The application of soil and water research has been hampered by imperfect understanding of some soil and water processes, and more often, by inadequate understanding of how different processes are connected. Some processes, such as the movement of water over and through soil, are reasonably well understood; others, such as the effect of soil loss on nutrient availability, or how commodity prices might affect conservation practices, are less well known. How a farmer's choice of agricultural practice is influenced to degrade or protect resources is even less understood. Such factors as land tenure, food-pricing policies, availability of inputs, as well as social and cultural constraints, affect the technologies a farmer will employ at any particular time. Little is known of the reasons a farmer places long-term stewardship or good husbandry of the land as a priority over immediate, short-range returns, particularly when living near the subsistence level. This imperfect knowledge base means that predicting the consequences of interventions is more art than science. Analysis of the results of interventions is thus always needed.

Data bases are a vital component of a valid systems approach, though multiple kinds of data can help fulfill this need. Although many fundamental processes operate within well-defined rules and knowledge can be extrapolated from one area to another, information on soil and water characteristics are site specific. To foster wider use of the systems approach, an international network needs to address standardized data base management and develop a set of common procedures to derive particular elements of the data base.

The need to bring a broader understanding of processes together with the site-specific data is a key component of the framework. This can be accomplished in a variety of ways and will often include the intermingling of scientific and local understanding of data obtained through standard, more universal methods with those generated at local levels through more traditional means.

Research in this framework, then, focuses on the need for understanding the local system and for evaluation of the relevant problems within that system. It is then possible to predict the effect of interventions and undertake means to control unwanted consequences, that is, to ensure sustainability at present or higher levels of productivity.

The scope of such research must pass through various scales of analysis, from the plant and soil interactions through local ecosystems to the meso scale (river basin size) system, for example. This diversity of scale is necessary to analyze sustainability at the crop or crop-mix level, at the level of the farm and its surrounding land and water, and at the regional level where management of watersheds and hill slopes can have a major impact on the other parts of the system. Recent advances in Geographic Information Systems (GIS) permit analysis of such spatially distributed phenomena. GIS enables data to be integrated through a common geographical frame of reference and fosters interdisciplinary research.

In past decades, the United States has been a leader in the production of research on the tropical soil, biotic, and hydrologic systems. Much of the research has been focused on theoretical problems related to components of these various systems, and has been driven by particular institutional and personal preferences. Consequently, much remains to be done in refining the research effort to deal with the real problems faced by tropical areas. When we add the goal of sustainability and realize that many of the tough problems of sustainability are those associated with more marginal lands, there is an even greater need to refocus effort, particularly in light of the poor record of sustainability in better endowed regions.

To be successful, future soil and water research must be organized in a systematic context. A driving force for that research will be the need of the peoples of the area and of the adaptability of the physical system to meet those needs. While the next chapter will outline important priorities that need attention, our overall priority will be to create a new and revitalized approach to the research process. Clearly, this will be a significant challenge, but it is attainable.

THE RESEARCH FRAMEWORK AND THE FARMER

To focus research more effectively on the needs of farmers in specific socioeconomic settings, researchers will need to spend more time on farms and in actual farming situations. They will need to understand the farmer's problems and the myriad variables the farmer considers during decision making. For instance, the farmer may not need a higher yielding corn, but rather one that is more resistant to moisture stress early in the growing season. Or perhaps the farmer most needs a variety that will compete effectively when grown within a suite of crops. The results of research need to be evaluated in the field under realistic conditions. This approach does not imply that component research is no longer needed, only that investigators must become more familiar with the wide range of cropping strategies employed by farmers, and with the pressing limitations that matter most to them. It is appropriate to mention here that many farming and land systems are managed by women and that research efforts and extension may need to be analyzed in light of gender relations.

Many of the agricultural systems in use in developing countries are complex, diverse, and changing (Altieri and Hecht, 1990). Researchers can benefit from understanding some of the common characteristics these agricultural systems share (OTA, 1988). For example:

- Many farmers use techniques designed to minimize risk, even if this means they obtain less than maximum yields;
- They rely primarily on local, indigenous knowledge, although new crops are regularly incorporated into existing systems;
- They often depend on biological processes and renewable resources in lieu of capital-intensive inputs;
- They commonly involve low cash costs but require relatively high amounts of labor; and
- Their management strategies are adapted to local cultures and environments, although social as well as ecological systems are showing increased problems because of mounting stresses.

No advance in either soil or water management practices can be useful unless the practice is adopted by the land users. Researchers, of course, have little or no control over nontechnical aspects of this critical decision-making process. But researchers can encourage the right choices. Indeed, researchers should try to develop a rich variety of innovations together with farmers, on the assumption that it is the farmer who is best able to choose from a set of alternatives one that is at once desirable and practical. Special attention needs to be given to devising incentives that encourage farmers to choose sustainable approaches. Enabling both land resource users and policymakers to exercise choice may well be the key to technology adoption.

One of the critical issues in sustainable natural resource management is to develop mechanisms that allow farmers to earn enough income so they can afford to strive for long-term sustainability instead of short-term gains during any one cropping season. Farmers do have a sense of stewardship and husbandry for the land; it is a challenge for the research and development establishments to find ways to harness this stewardship for the long-term good of the farm and the farming community. Only well-integrated activities involving several disciplines, including the social and biological scientists working with the farmers, can bring the more holistic approach needed to implement and maintain more sustainable agricultural strategies.

3

Research Priorities To Support Sustainable Agriculture

Agriculture in tropical environments faces different constraints than in temperate regions, and this affects soil and water research needs. Six broad areas merit priority attention: overcoming institutional constraints on resource conservation, enhancing soil biological processes, managing soil properties, improving water resource management, matching crops to environments, and effectively incorporating social and cultural dimensions into research. In addition, better use of indigenous knowledge and improved communications can enhance the implementation of research results.

Two critical indicators of deterioration in agricultural systems are declines in the quality of soil or water. Poor management of either of these resources quickly leads to decreases in farm productivity. Thus there is an urgent and ongoing need for research to devise ways to manage soil and water resources more sustainably.

A large proportion of the world's developing countries is located in tropical environments.[1] Tropical agricultural systems differ from temperate systems in significant ways, both physically and institutionally. Several of the unique characteristics of tropical environments—from mountain highlands to arid rangelands to humid forests—played particularly important roles in the

---

[1]Much of the developing world is located in Africa, Asia, and South and Central America. These are essentially tropical continents—that is, most of the land area is located either in tropical latitudes or is influenced by tropical atmospheric systems. Within these continental areas, there are several high mountain areas and plateau areas that are characterized by distinctly nontropical climates. The generalizations about tropical environments do not apply to these upland zones, which need a much more specialized regional approach to their particular soil and water systems than is possible in this overview document. It is essential to acknowledge that the developing world exhibits a tremendous diversity of environments and cultures.

committee's determination of the major issues needing research. Some of the characteristics that make tropical environments especially challenging include the following:

• Lack of a cold season or frost, which in other climates brings a break in production, thus affecting pests, diseases, and moisture levels;
• Variable timing and duration of water supplies in both dry and wet regions, creating severe moisture stresses;
• Year-round growing seasons in some wet areas, with effects on crops and pests as well as accelerated leaching of nutrients;
• Greater biological diversity than temperate environments, and thus greater diversity of crops, soil organisms, and pests;
• Highly weathered soils, and in some places very young soils;
• Shortages of fossil fuel and other capital-intensive inputs; and
• Significantly different social and institutional contexts and traditions.

In the tropics, especially where access to fossil fuel resources is limited, management strategies must be more biological in nature and must rely on the use of appropriate choices of germplasm, cropping systems, and techniques to fit specific ecological niches. Many methods employed in the tropics were developed over time through trial and error; they vary greatly among different geographic regions and cultures. Often, attempts to transfer research strategies elaborated in temperate zones into tropical environments have failed to recognize these fundamental differences. At the same time, the diverse social institutions, kinship patterns, resource access, and tenure relationships in developing countries do not necessarily operate in the same manner, or respond to the same logic, as parallel structures in the industrialized world. Indigenous, colonial, and modern patterns of resource access and regulation can, for example, operate simultaneously on a given water resource or piece of land, although often with conflict. Recognition of these basic differences between tropical and temperate agriculture must be a main factor in the selection of priority areas for research.

Given the problems faced by tropical agriculture, the unique characteristics of the environments and cultures, and the strengths and weaknesses of the existing knowledge base, research in the following areas could offer great rewards in support of sustainable agriculture and natural resource management:

• Overcoming institutional constraints on resource conservation;
• Enhancing soil biological processes;
• Managing soil properties;
• Improving water resource management;
• Matching crops to environments; and
• Incorporating social and cultural dimensions into research.

A wealth of time-tested indigenous knowledge exists and this human resource should be tapped to support these goals. Special potential lies in the blending of traditional methods with modern innovations. One of the most intractable problems yet to be faced is the difficulty of communicating new ideas to the farmer. Research and development organizations have struggled with this problem for many years, and it remains a high priority issue.

The committee faced a difficult decision when deciding how to present these research priorities. On the one hand, it wished to stress the need for interdisciplinary research, particularly research that integrates technical and institutional dimensions. Such research often does not fit neatly into categories. On the other hand, the committee needed to use some organizing structure to make the ideas accessible and useful to the research community. Such structures often imply boundaries and separateness, however, and they diminish the importance of interrelationships. In the end, while the committee opted to present its priorities in six distinct categories, it wishes to stress the need to move beyond compartmentalized thinking and toward more integrated approaches (see chapter 4 for more discussion).

The following discussion explores the six priority research areas and gives examples of the types of investigations most needed. Within each, some examples of research fields are listed. These are divided into two categories: critical research priorities and other priority topics.

In defining research needs, the committee considered and set aside many potentially valuable areas of research. For instance, further research on ground water modeling and soil physics is not recommended, largely because our knowledge in those areas already is extensive. Research on the use of mycorrhizal fungi is not highlighted because it lacked immediate, practical value to the farmers in most need of sustainable agricultural strategies. Research on soil acidity and its effects on the growth of agriculturally important crops is not listed because this is reasonably well understood and many measures are already known to correct such problems. Likewise, nutrient requirements for most crops can be fairly well predicted, and the fundamentals of how to meet these needs have been extensively researched. The primary deficiency in our decision-making is the limited ability to specify appropriate management practices that will be socially and economically acceptable for site-specific conditions. Thus, while the emphasis of the research topics outlined here is less on individual components of soil and water systems and more on the broader, and variable, systems themselves, the need for location- and culture-specific adaptations of these general approaches is of course implied.

OVERCOMING INSTITUTIONAL CONSTRAINTS
ON RESOURCE CONSERVATION

Resource conservation is at the heart of all efforts to develop and implement more sustainable approaches to agriculture. It is increasingly clear, however, that resource conservation must be understood to include more than the technical approaches. Sustainable agriculture cannot be obtained without attention to the economic, policy, and institutional elements of resource conservation. Although a large part of the focus on sustainability is directed at the farm level, it is obvious that soil and water management issues relate to the management of the whole environment.

Soil and water management are affected by many factors external to the farm, such as the pricing policies of national or international bodies, tenure rights, or available labor. Institutional constraints on resource conservation are as critical to the sustainability of agroecosystems as on-the-ground soil and water management techniques.

Critical research priorities should include:

• Studies of land and water resource tenure and access policies that affect long-term stewardship and sustainable agricultural practices;

• Analyses of social, political, and economic dynamics of pricing policies and how these affect the stewardship or degradation of land and water resources; and

• Thorough evaluations of in situ and ex situ germplasm conservation, and the relative merits and problems each offers for maintaining biodiversity to provide a wide range of genetic options for varying hydrologic regimes and soil fertilities.

Other priority issues should include:

• Evaluation of local and regional institutional arrangements to improve the integrated conservation of soil and water;

• Evaluation of short-term incentives that might be used to implement long-range sustainability goals; and

• Assessment of the nature and impact of national policies that affect the use of industrial inputs such as fertilizers, pesticides, and mechanization.

Both on-farm and off-farm resources affect resource management. On-farm factors such as labor availability, land tenure, and access to resources are influenced by larger scale economic and political forces that affect the choices that farm households and communities make about their resource use. On the farm, research should continue to address how sound stewardship can be used to meet the needs of both the farmer and society. However, off-farm questions, historically, have not received as much attention, especially as regards implementation of research findings.

Resource conservation is a clear example of an issue that should unite agriculturalists, soil and water managers, and environmental scientists concerned with the long-term use and protection of the natural resources base. The sometimes adversarial relationship between these interests is counterproductive. Instead, all should be attempting to integrate their special knowledge in pursuit of answers to the complex local, regional, and institutional constraints on resource conservation.

ENHANCING SOIL BIOLOGICAL PROCESSES

Sustainable agriculture requires maintenance of the soil biota, particularly in areas of low-income, low-input agricultural systems where soil biological processes are critical for sustaining and enhancing soil fertility. Soil structure, nitrogen fixation, nutrient availability, and control of soil-borne pathogens and pests all can be manipulated by organic inputs and vegetation management.

In most soils, high levels of soil organic matter are essential for good soil structure. Yet many important questions remain regarding the avail-

Improved methods of using soil biological processes to best advantage are particularly important in areas with low-income, low-input agricultural systems. Researchers at this field station in Senegal are investigating root nodulation on fast-growing trees; the AID-funded project involves Senegalese, French, and U.S. scientists. *Credit*: Michael McD. Dow, National Research Council.

ability and management of organic matter. For instance, the release of plant nutrients through rapid decomposition is a key way to optimize crop yields, but more research is needed to understand the role of macrofauna in this process. Recent work on the manipulation of earthworms and termites appears promising. The benefits of organic matter in the maintenance of soil quality and crop productivity have long been well known, but only recently have researchers started to elucidate the role of organic matter additions in variable charge soils on soil acidity, the ability of soils to hold nutrients, and phytotoxicity.

Another example of a promising research area is biological nitrogen fixation (BNF). Although the general process involved is well known, as evidenced in farmers' wide use of legumes, more needs to be known about its ecology—including plant-microbe-climate-soil interactions. A better understanding of these interactions would facilitate our capability to manage these processes. For instance, successful use of BNF in cropping systems is often erratic and unpredictable, and research could develop methods for BNF inoculation at different sites. New methodologies also need to be developed for quantifying BNF by trees, since current measurement techniques are difficult in the field and often inaccurate.

Agroforestry systems are widely and traditionally used in many tropical areas. These practices are becoming increasingly important in agriculture for many marginal areas, especially the wetter tropics, and in areas where fuelwood is scarce. Nevertheless, little scientific data exists on the soil impacts of agroforestry elements such as rooting patterns, allelopathy, nutrient cycling, and the ability of some species to competitively take up or accumulate scarce nutrients. The complex manner in which tree products are integrated into household, local, and regional economies needs to be assessed as well. Issues related to the constraints imposed by limited human labor also require attention.

Effective management of soil biological processes to cycle and fix essential crop nutrients can reduce expenditures on fossil fuels. The synchronization of nutrient release from organic inputs to meet nutrient uptake demand by crops is critical and little understood. Often, nutrients are released in the decomposition process when the plant does not need them, and subsequently may be lost. The use of cover crops, mulches, crop rotations, and minimum tillage are known to control many soil pathogens and nematodes and maintain soil tilth. More information is needed on the actual mechanisms involved.

Although they vary by region, many sustainable agricultural systems in the tropics include livestock. Their role includes the consumption of crop residues and the production of organic matter and nutrients as well as the control of unwanted vegetation. Work on input-output or recycling models would help many small farm enterprises.

ORGANIC RESIDUES AND SOIL ACIDITY PROBLEMS

Soil acidity is a major problem in soil management, and the use of organic residues is a promising potential tool for farmers in the tropics. As commonly used, the term "organic matter" is not the same as "organic material." Organic matter, in the conventional sense, refers to the well-decomposed, origin-unrecognizable organic portion of the soil; organic materials include more intact organic residues such as plant residues, green manures, mulches, and composts. The potential role of organic material varies with its composition and the soil to which it is to be applied. Organic material additions can alleviate the low calcium, magnesium, and potassium conditions characteristic of acid soils because it contains relatively large amounts of these nutrients.

Much remains to be learned about the potential role of organic residues in alleviating soil-acidity problems. Research could help fill some of the information gaps regarding the use of organic material to combat acidity. For instance, what organic materials and decomposition products are most effective in reducing aluminum content and solubility in soils? What are the effects of differing surface mineralogy on organic material and its decomposition products? What is the duration of the liming effect of organic materials? Is there an optimum quantity of organic material to add to the soil? In addition to the chemical and physical aspects of the practice, an important economic consideration is whether there is sufficient organic material available for application in the near proximity of the farmer's field, and, if not, what are the costs of labor and transportation to move large quantities of the material from adjoining land.

One important variable to be considered when predicting the improvements possible through organic residue management is whether the organic material was grown in situ or obtained from an exterior location. If imported, the nutrient content of the organic material is contributed to the soil system. If it is grown in situ, the overall benefit is usually less because the nutrients are simply recycled. In some cases, however, recycling and bringing nutrients from deep zones in the soil profile can substantially improve the surface soil—the root zone for most annual food crops.

The role of organic material in reducing aluminum toxicity, often the most detrimental aspect of the soil-acidity syndrome, includes the chemical complexing of the aluminum in solution, thereby reducing its activity. Organic material additions can, in some cases, also alleviate phosphorus deficiency in acid soils by supplying phosphorus directly, by reducing phosphorus sorption capacity, and by complexing soluble aluminum and iron, thereby increasing soluble phosphate concentrations. In large amounts, organic materials can reduce acidity simply by increasing the soil pH. In smaller amounts, the type of organic material becomes important, for example cowpea is more effective in reducing acidity than leucaena.

Critical research priorities should include:

• The ecology of biological nitrogen fixation (BNF), including plant-microbe-climate-soil interactions and improved methodologies for predicting BNF response and for quantifying BNF by trees;
• Analysis of agroforestry systems, including rooting patterns, allelopathy, nutrient accumulation, and nutrient cycling;
• The role of organic matter in variable charge soils on soil acidity, phytotoxicity, and similar factors; and
• Soil process-related management techniques to control soil pathogens and nematodes (e.g., crops rotations, cover crops, soil organic matter manipulations).

Other priority issues should include:

• The role of macrofauna in soil fertility;
• The effects of minimum tillage on soil biota, including pests and animals;
• The role of livestock in small farm systems; and
• The enhancement and maintenance of biotic inputs for sustainability in low industrial input systems, including synchronization of nutrient release from organic inputs to meet nutrient uptake demand by crops and appropriate biotechnology efforts.

MANAGING SOIL PROPERTIES

Rarely do the inherent properties of the soil provide an ideal environment for agricultural use. Fortunately, many of the limitations are amenable to improvement through inputs, manipulation, and other management practices. Research has led to substantial progress in identifying the fundamental constraints and basic principles of soil management, although such work has been conducted largely in developed countries in temperate regions. However, for both the developed and developing world, a central weakness is our limited capability to provide optimal site-specific soil and water management practices that can be employed by individual land users within the context of their needs and the prevailing social, economic, and political climate. Thus, this ability to translate scientific knowledge about soil characteristics and plant growth into useful information for farmers is a major research need for the future. Better management of the chemical and physical characteristics of the soil is critical to sustainability.

Critical research priorities should include:

• Efficiency in use of organic materials;
• Sources of nutrient amendments;

- Mechanisms and amelioration of soil compaction and crusting; and
- Strategies for restoring degraded lands.

Other priority issues should include:

- Soil loss and its effects under different management strategies; and
- Improvement and use of diagnostic technology for nutrient availability.

Research on better management of soil properties will involve both the chemical and physical characteristics of the soil system, and this area offers special potential in the context of the limited use of capital-intensive input characteristics of many developing countries. For instance, work on sources of nutrient amendments is key because low residual levels of essential elements is a common cause of soil infertility in the tropics. This condition usually can be corrected through amendments. Much of the technology for these types of management practices has been developed in areas where purchased inputs are readily available. In developing countries, however, the capital for such investments and the managerial capability to deal with the type and level of technology is limited. Therefore, alternative practices should be provided that are compatible with local natural resources and social, cultural, and economic conditions. Special emphasis should focus on biotic amendments.

Similarly, research on the efficient use of organic materials is critical. Organic materials can have multiple benefits in reducing or alleviating many soil chemical problems. Notable among these are providing nitrogen and other essential nutrients and correcting soil acidity. The technology for effective and efficient use of organic materials—for example, nutrient-accumulating species of plants and management of residue—is not available in a form suitable for most of the developing world. This is an area where the blending of scientific knowledge with indigenous knowledge offers great potential benefits.

Mechanisms to ameliorate soil compaction and crusting are important because these frequently lead to decreased water infiltration, increased runoff, increased erosion, and reduced stand and growth of seedlings. General principles for dealing with these problems are reasonably well understood, but management practices that would be useful in the developing world require better knowledge of fundamental causes and alternative solutions.

Given the ever-increasing pressures for production in the developing world, strategies for restoring degraded lands will also prove key over the long term. Past mismanagement has resulted in the abandonment of extensive areas. In the Amazon basin, for example, roughly half the area that has been cleared—some 6 to 7 million hectares—has been abandoned (Toledo and Serrão, 1982). The basic problems stem from a variety of causes, both chemical—such as loss of fertility and high soil acidity—or physical—such

A donkey-drawn sandfighter, shown here at the ICRISAT Sahelian research center near Niamey, Niger, is one technique for managing and improving soil properties. The sandfighter is used soon after a rain, when its tines can dig the damp sand into shallow depressions and small, tight clods. The broken surface traps windblown sand, reducing erosion and protecting young crops from sand blast and burial. *Credit*: Neil Caudle, North Carolina State University.

as crusting, dispersion, or compaction. In many cases, application of lime-stone and phosphorus will ameliorate the chemical problems to the extent that the land can be used in an economically productive manner. But where the primary degradation problem is physical, remedial measures are more difficult and time consuming. The practices may require chemical treat-

RAINWATER HARVESTING

Rainwater harvesting is the practice of collecting precipitation for domestic or agricultural use. It has been employed in various forms for more than 4,000 years and in areas that range in annual rainfall from 20 mm to 1,800 mm. Collection schemes vary from clearing hillsides of rocks and gravel to increase runoff and direct it toward cultivated fields further down the slope to collecting water from rooftops in small impoundments.

Rainwater harvesting is primarily for small-scale use for farms, villages, and livestock. Approaches to rainwater harvesting vary greatly. Some practices rely on alteration of the land surface and require construction, such as the building of water catchments and tied ridges. Another approach, one particularly good for porous or unstable soil, is to cover the soil with a waterproof cover. Plastic sheeting, butyl rubber, and metal foil are low-cost alternatives for rainfall catchments. Gravel can be placed on top of plastic to protect against wind and sun damage. These catchments, if properly built and maintained, can have an expected useful life of more than 20 years.

Water harvesting has the potential to enhance food production in water-short semiarid and arid Third World countries. It is especially promising for developing countries because it provides water without requiring fuel or power. However, specific techniques will need to be developed to meet site-specific soil, climate, and socioeconomic conditions. Water harvesting schemes have the potential to be especially useful in areas with the following characteristics (NRC, 1974):

(1) Clay soils. On these sites much of the water runs off during rainstorms. Small ponds built in the local watershed can be used to harvest water during the rainy season and store it for use during the dry season for domestic uses and irrigation of food crops.

(2) Laterite toposequences. Many of these toposequences have impervious layers at the top that allow little or no vegetative growth. Thus rain falling on the impervious caps runs quickly down the slopes, causing erosion and exposing infertile subsoils. A series of check dams or levees can slow or stop the movement of water and store it temporarily for domestic or agricultural use. These check dams may also reduce erosion and make the soils down the toposequence more useful for crop production.

(3) Gravel mulches. One of the primary sources of water losses in semiarid environments is soil water evaporation. An established means to reduce such loss is by gravel mulches. In many areas, for example the Sahel, laterite gravel is abundant. Spreading this gravel over the surface can reduce water evaporation and thus increases the water available for human use.

ments, but more frequently involve long-term fallow periods with trees or cover crops as the primary vegetation. As with other soil management practices, restoration strategies must be tailored to individual sites and circumstances. The ability to make these site-specific recommendations remains a major challenge for research.

Soil degradation under different management strategies is important because the choice of cropping system can have a major influence on the loss or retention of soil. The challenge is to employ alternative systems that will enable the farmer to use the land in a manner that minimizes soil loss and damage. The option generally is not whether to use the land or not—circumstances often require it; it is the method of use that becomes the point at issue. On-site degradation is only one of the issues. Determining the effects of soil loss on off-site locations is increasingly important. The impact of erosion on downstream ecological and agricultural systems needs to be assessed. Acquiring a better understanding of the social and economic dimensions of soil degradation and providing incentives to the farmer for erosion control and prevention measures would go a long way toward enhancing land use. It is also essential to keep in mind the whole soil-water system and to conduct research that looks at the dynamic relationships between these critical elements.

Finally, improvements are needed in diagnostic technologies to measure nutrient availability so deficiencies of essential soil nutrients can be corrected more easily. Such corrective practices are expensive, whether achieved by purchased inputs or by organic residues. The cost and efficiency of remedial measures can be improved if specific and quantitative data are available on the prevailing level of the nutrient in the soil. Substantial progress has been made in diagnostic analyses of soils to serve as a guide for nutrient inputs. However, the applicability and adaptation of these techniques to developing countries has been given relatively less attention.

IMPROVING WATER RESOURCE MANAGEMENT

For much of the tropical world, water is the key natural resource, and managing variable, dynamic water supplies thus is a critical challenge if agriculture is to be sustainable. As populations grow and urban and industrial water demands increase, competition for water has intensified. Research must address a spectrum of issues ranging from rain-fed agriculture to irrigation, from the effectiveness of small-scale indigenous techniques to the impacts of large-scale impoundments. It must look at water's role in dynamic agricultural systems and, in particular, its close interrelationship to soil resources. It must move beyond the technical questions toward questions of how to apply technology in the diverse cultures and ecosystems of the developing world.

DEALING WITH THE UNCERTAINTIES OF EARLY RAINS

One fundamental difference between most temperate climatic zones and the tropics is that in temperate areas crops generally are planted in the spring in a water-saturated soil environment, while in most parts of the tropics with a dry season, crops have to be planted in dry ground or in newly moist ground at the beginning of the rainy season. There is pressure to plant early because the food is needed and the growing season is limited, but there is also a real danger of planting too soon— the first rains may not be substantial and the crops may wither. Planting too late, in turn, can present other problems. Wet soil is hard to work, and pests start to cause problems.

Problems associated with the vagaries of rainfall in the tropics may be reduced by selectively breeding crops to withstand early drought, by developing better predictions of the pattern of rainfall in a season and ways to communicate this information to farmers, and by developing transplant techniques for crops that have not traditionally been transplanted. Also, soil-imprinting techniques that act to concentrate sparse rainfall at the base of each seedling are helpful. All these areas need research.

The use of rice nurseries before the onset of monsoon rains is certainly one of the oldest strategies to avoid water stress in the seedling stage. Rice is planted, irrigated, and fertilized in small nurseries at the end of the dry season. When the rains start, and fields are sufficiently flooded, rice is transplanted. Other similar examples may be found in the savanna areas of Africa where sorghum and corn are planted in small irrigated nurseries, and then transplanted to larger fields once the rains have commenced in earnest. The use of soil imprinting and tied ridges also can help to concentrate moisture at the base of seedlings.

Four general areas merit attention: (1) techniques for water capture and impoundment, with attention to indigenous, small-scale techniques; (2) strategies to enhance water conservation so maximum return is gained from each drop of water; (3) methods to reduce irrigation-related soil degradation, such as salinization; and (4) larger-scale approaches for watershed and landscape management. Some of the greatest potential benefits of improved water management may well be found "at the margins" of the water management field, such as the improvement of little known or innovative approaches and technologies. The challenge is to help farmers optimize their use of available water while maintaining the quality.

Critical research priorities should include:

• Developing techniques to help farmers plant and maintain crops during the uncertain, early stages of the rainy season. These might include studies of seedling resistance to the vagaries of water supply, transplanting methods, and strategies to provide more secure environments early in the cropping cycle.

• Describing and evaluating the array of existing water-harvesting techniques, particularly indigenous ones, and consider their possible effectiveness in new environments.

• Investigating the role of aquaculture in farming systems and especially in irrigation systems. Are there ways to combine irrigation and aquaculture in regions where this has not been a tradition?

• Investigating techniques for making the best combined use of surface and sub-surface water in irrigation and dealing effectively with drainage, including an analysis of indigenous technologies.

• Investigating the policy and political issues of pricing and subsidies of

Innovative thinking can help surmount the constraints posed by insufficient rain early in the growing season. In Nigeria, local farmers have developed a system where sorghum is started in irrigated nurseries and then transplanted into the fields once the rainy season is under way. *Credit*: Hugh Popenoe, University of Florida, Gainesville.

Many types of microcatchments are being developed to capture and channel limited or sporadic rainfall to crops. In the Negev Desert of Israel, this microcatchment concentrates water around an almond tree. Its design was modeled on evidence of similar structures discovered during archeological research. *Credit*: Mike Austin, University of Hawaii.

water and products in water-scarce areas and the long-term impacts these have on agricultural strategies and soil management.

Other priority issues should include:

• Conducting comparative institutional analyses of risks and benefits of water management strategies, including local as well as large-scale institutions. This would include attention to indigenous systems of water access and tenure.

• Analyzing the economic, social, and environmental effects of irrigation at various scales. There has been a plethora of studies of large irrigation projects and their problems, but much less attention on the smaller systems and on supplementation strategies. Can modern systems be adapted to augment various indigenous systems and vice-versa?

• Investigating the waterborne diseases and weed infestations associated with canals and drains for irrigation development and techniques to combat these problems.

• Analyzing adaptations to short-term drought within the cropping season.

Fortunately, much of the knowledge about physically manipulating and managing water resources for both rain-fed and irrigated agriculture is more or less generically transferable, although the scales at issue and the socioeconomic impacts are quite diverse. However, large gaps in our knowledge base still remain, especially in the areas of agroforestry, mixed cropping systems, and sustainable low-input rain-fed agriculture. Current knowledge focuses on full reliance on irrigation, while many questions about supplemental irrigation remain unaddressed. In addition, our knowledge base is still severely lacking in areas related to the social and institutional aspects of managing public or village-level irrigation enterprises and dealing with the questions of conjunctive use.

How policy and institutions affect the allocation and use of water resources, and thus sustainability, are complex but especially important questions. For instance, what incentives will encourage maintenance and repair of water management systems? What risks do farmers face—whether hydrologic, climatic, economic, or political—and how do farmers make decisions in light of these risks? What are the equity implications of particular interventions? Since both rain-fed and irrigated agricultural systems often produce significant environmental impacts, research is needed to evaluate and improve the criteria and methods for carrying out environmental impact studies for project design and existing projects.

MATCHING CROPS TO ENVIRONMENTS

Research to enhance the matches between crops and environments is an ongoing and still critical endeavor. This work should include studies focused on the concept of ecological niche and the ways in which agriculture might benefit from use of appropriate organisms, including opportunities offered by genetic manipulation. It also means more effort is needed to match crops and production strategies to social, cultural, and economic environments. Although much research remains to be done regarding alterations of the environment to suit crops, for sustainable agriculture the focus should be on selecting and adapting crops and management choices to various settings.

Several processes are necessary to match crops to environment. First, the environment itself must be characterized in terms of soil parameters like salinity, alkalinity, aluminum toxicity, nutrient deficiencies, slope, and soil erodibility. Next, climatic features, including temperature, photoperiod, insolation, and availability and seasonality of water, must also be incorporated. Local vegetation types, including classification categories of local people, should be described. The broad varieties of crop plants themselves, including unusual genotypes of existing crops, merit description. As these are the outcome of manipulation of crosses and planting materials selected

MATCHING CROPS TO ENVIRONMENTS: THE WINGED BEAN

The story of the winged bean provides an example of successfully matching crops to environments. This tropical legume was a little-known crop when the National Research Council (NRC) published a small report on the plant in 1975. At the time, there were only 12 known varieties, with most of those grown in Papua New Guinea and Southeast Asia. But the plant showed exceptional promise—its seeds had a nutritional value almost the same as soybean and, unlike the better known soybean, it thrived on poor, acidic soils in the humid tropics, tolerated high rainfall conditions, and was comparatively resistant to pests and diseases.

By the time the NRC published a second edition of the report (NRC, 1981b), perhaps as many as 600 varieties had been collected; a worldwide information exchange network was in place; researchers in more than 50 countries received a newsletter, *The Winged Bean Flyer*; and research trials were under way at sites around the world. Ten million dollars had been invested in winged bean production in Thailand alone and 6,000 acres had been planted in the Ivory Coast. Few crops have risen so quickly out of obscurity.

The winged bean's rapid success comes in part because it meets a real need: people of the hot, humid tropics need better plant sources of protein. Sometimes described as "a supermarket on a stalk," winged bean meets this need and more. In addition to its soybean-like seeds, its leaves are rich in vitamin A and can be cooked and eaten like spinach; its shoots resemble asparagus; its flowers, when steamed or fried, make a mushroom-like garnish; and its tuberous roots are like nutty-flavored potatoes with twice the protein.

But what really turned potential into success was a concerted research effort. The knowledge about the crop was assessed, gaps identified, and its potential investigated. A strategy for international research and testing was developed and pursued. All in all, the winged bean offers a concrete example of the benefits possible through the blending of indigenous knowledge and scientific method when it comes to looking for crops to suit particular ecological niches.

over time to tolerate local constraints, they have particular merit in guiding sustainable agriculture research efforts.

To comprehend underlying cultural, economic, and social context from the household level, two approaches are required. The first involves the scientific quantification of ecosystem parameters by standard analytic methods. The second involves the mobilization of local knowledge of landscapes, soils, weather, water cycles, and planting material, which can add subtlety and depth that those outside the scientific community cannot readily see. One potential research area involves analyses of the processes of adaptation, and the emergence of new cultivars through the assiduous collection of land races, conventional plant breeding, and other forms of biological manipulation.

In general, the matching of crops to tropical environments has not received adequate research attention because of a traditional emphasis on "conventional crops," annual crops, and monoculture production systems. Critical research priorities should include:

• Improved understanding of indigenous knowledge, especially local soil and crop taxonomies as well as the analysis of the array of environmental knowledge that could help orient current research;
• The potentials and problems of erosion-prevention crops and strategies;
• Crops developed and selected for their adaptations to chemical stresses such as aluminum toxicity, alkalinity, and soil and irrigation water salinity;
• More analysis of mixed management practices, including techniques such as alley cropping, agroforestry, and successional management, and their potential to meet multiple needs including, but going beyond, yield.

Other priority issues should include:

• A typology of environmental parameters for different crops, including attention to climatic constraints, microclimates, and temporal changes.

Local peoples tend to manage a diversity of agroecosystems and natural resources. Agriculture in many societies is not conceptually limited to what the developed world calls "crops"; instead, agriculture often includes the management of a variety of semi-domesticates, weeds, forests, wildlife, and other elements of the local environment. Indigenous management of landscapes is often difficult to see.

The management of natural resources in this broad sense may have a diversity of goals beyond increased yields. Thus, low-yielding but highly reliable or low-risk varieties, ceremonial cultivars, or stress-tolerant varieties, personal favorites, starvation foods, and particular genetic lines may all be included in an agricultural strategy. These "other logics" may be overlooked or dismissed by those with only a yield-oriented outlook, but they are essential to any research agenda that focuses on land areas with serious water, soil, and income constraints or erratic and less predictable climate and economic milieus. Attention to such strategies, of course, should provide insights into long-term sustainability of regional production systems.

Experimental evaluations of indigenous knowledge of land-vegetation management are scarce because these systems are too dynamic for most experimental agronomic research design. Still, the elements and principles embodied in such systems can provide hypotheses for later testing and can serve as a springboard for forming agronomic strategies that enhance soil and water management in marginal or complex environments. For instance, such information could be particularly useful in identifying and disseminating plants that are effective for erosion control (for example, vetiver grass). These efforts can also provide new candidates for soil-nutrient management in various kinds of agroforestry and successional systems. If sustainability is to be incorporated into the development agenda, the relevant protective contributions of the elements of agricultural systems and their effectiveness must be evaluated and given weight in any economic and social evaluation. This information can provide insight into general principles with ecosystem or regional relevance.

In matching crops to environments, large potential may exist for transferring valuable or useful plants from one ecosystem niche to other analogous sites. The winged bean is one of the better known cases, but many lesser known species could undergo similar development processes. The National Research Council, through its Board on Science and Technology for International Development, has published a series of books that seek to bring attention to underexploited crops with potential value for specific ecologi-

Amaranth is one of many lesser known crops with significant potential in sustainable agricultural strategies. These researchers are on a field trip looking at different varieties of amaranth in Cusco, Peru, at the Center for Andean Crops Research. *Credit*: Michael P. Greene, National Research Council.

cal niches. For instance, reports on jojoba (NRC, 1985), saline agriculture (NRC, 1989c), and amaranth (NRC, 1983a) highlight crops for dry regions. The important role of livestock has been described in books on water buffalo (NRC, 1981a), little-known asian animals (NRC, 1983d), and one on microlivestock (NRC, 1991a). Valuable forestry options are examined in reports on calliandra (NRC, 1983b), casuarinas (NRC, 1983c), and leucaena (NRC, 1984). Many more crops and livestock, not widely known, show similar promise and could help to enhance the productivity of agriculture, particularly in marginal environments.

INCORPORATING SOCIAL AND CULTURAL DIMENSIONS

Most people involved in research and development for developing countries now agree that research—if it is to be of practical value—must incorporate social, cultural, and economic factors as well as technical and scientific ones. Just as a particular soil management strategy must match a particular soil, so must it match the people who will be responsible for implementing it. But actually accomplishing this goal is difficult. Agricultural scientists often lack the training, experience, and time to integrate such factors into their research (Colfer, 1987).

Actively incorporating these dimensions into research and its applications requires a shift in the model common in developed country agriculture, where the researcher does the science and passes the product to the extension agent, who, in turn, conveys the information to the farmer. This unidirectional approach was successful in the United States in part because the scientists and extension agents involved typically had farm backgrounds— they instinctively knew the social, cultural, and economic constraints at issue because the environment (both biotic and social) was a familiar one. As the distance between farmer and researcher grows, however, the greater is the need for a different model of communication. The knowledge, experience, and capabilities of the local people must be incorporated into the research process so it is, in essence, a joint endeavor (Colfer, 1987). A collaborative approach is necessary, with mechanisms for feedback and refinement of the research plan. Although this need for a more interactive approach to research has been discussed widely and for some time, it is still a significant constraint on the effectiveness of research in support of agriculture, particularly efforts to develop sustainable agricultural strategies. Consequently, the committee wishes to emphasize the need for serious, continued attention to the social and cultural dimensions of research pursuits. In the end, this may be one of our most important findings.

It is no longer appropriate to relegate these human dimensions of research to secondary, separate status. A soils project, for instance, should emphasize what people do to the soil at the site, and what other factors affect that action. The goal is to identify opportunities for constructive improvement in the existing system. The social and cultural dimensions of the research must be kept focused on issues of relevance to the other scientists involved, and on overall project goals. At most sites, important areas for research will include:

• Indigenous soil, water, and agricultural classification systems;
• Allocation of time and division of labor both by gender and seasonality;
• Land tenure and access to resources;
• Subsistence, nutrition, and the cash economy; and
• Values, indigenous views on the people-land balance, and attitudes toward agriculture, soil, water, children, and the future.

4

Supporting an Integrated Research Strategy

An integrated research strategy requires institutional mechanisms and structures that link research organizations with clients, and also with the different components of research. Additional mechanisms are needed to reassess research priorities on an ongoing basis and to create locally oriented data about soil and water resources.

A collaborative, integrated research strategy will not develop without a conscious effort on a number of fronts. Changes are needed in the mechanisms and structures used by research organizations and funding agencies. Mechanisms need to be in place for various reasons:

- To link clients with researchers;
- To link components of research;
- To evaluate the effects of research;
- To set priorities as an ongoing task; and
- To deal with the problem of obtaining site-specific data to help solve site-specific problems.

STRUCTURES TO LINK USERS AND RESEARCHERS

A basic issue in targeting research to the needs of users is the pattern of communication and feedback. These chains of communication can be complex. Because a great deal of research relevant to the tropical context is conducted outside the areas of application, researchers in universities and in international organizations need to be linked with the users of the research in effective ways. Traditionally, crop research went through a hierarchial sequence from basic research to field testing to promotion by extension agents. But this structure does not work well in many developing countries and is not adaptable to soil and water management issues. Intermediate

NETWORKS IN INTERNATIONAL AGRICULTURAL RESEARCH

Viable research networks can play an essential part in solving many of the research problems. What is a "network" and what makes one successful? A 1990 book, "Networking in International Agricultural Research" provides a comprehensive review of networks (Plucknett, D. L. et al). "Networking," the authors note, "is a new name for an ancient practice." They define a network as "an association of independent individuals or institutions with a shared purpose or goal, whose members contribute resources and participate in two-way exchanges or communications." Networks are generally decentralized and lacking a well defined hierarchy of authority.

Networking has a long history in the sciences as a mechanism for sharing research results and information. It is especially well developed in agricultural research, and promises increased efficiency. It multiplies the effectiveness of individual scientists or institutions, encouraging diverse talents and expertise to be applied to problems. "Networking is no panacea for lagging agricultural production or inefficient agricultural research but it can be a powerful way of improving the quality and impact of research."

agents who can transfer research findings to the field and who can also communicate farmers' needs to the researchers, increasingly important in the process of research and technology transfer. Private voluntary organizations and cooperative host country universities are institutions that are particularly important in this regard. Special efforts are required to encourage networks that link institutions in a two-way exchange involving researchers, research institutions, universities in host countries, private voluntary organizations, cooperatives, and farmers.

Many of the existing research structures will need to be modified to support the needs of sustainable agriculture and natural resource management. Currently, much agricultural research still is done within the confines of disciplinary boundaries or as dictated and supported by various commodity groups and state funding sources. The reward system for the researcher is based on some of the same constraints; for instance, professional recognition is tightly bound to within disciplines. Little acknowledgement is available for a researcher involved in interdisciplinary efforts. Trends away from rural sociology and cultural geography within the social sciences have provided fewer interfaces with agricultural disciplines on rural development problems. Likewise, efforts to bridge the natural resource disciplines and agricultural disciplines at universities have been weak at best, and almost adversarial in some cases. New efforts should be made to change the administrative configurations of diverse groups of scientists working on common problems. The institutional reward system should reflect these changes. Rewritten mission statements, goals, and objectives would contribute to harmony among agriculturists, natural scientists, and social scientists working on common strategies for sustainable development. This new vision could help lead to more responsive technologies.

In summary, linking clients with researchers will entail a sensitive restructuring of the mechanisms and procedures now used by the research establishment. It is, however, the sense of the committee that awareness of this need has grown rapidly in the last few years and ways of enhancing better communication between researchers and research users will find a ready audience.

STRUCTURES TO LINK RESEARCH COMPONENTS

A second need is for the establishment of networks to facilitate the integration of different research components and to then focus these efforts on problems identified in the field. To be successful, such structures would require a strong client-driven component and assigned resources to facilitate the process. Many networks now exist that support the component approach to agricultural research in the tropics. Some of the better known ones focus on fast-growing nitrogen-fixing trees, biological nitrogen fixation, the winged bean, amaranth, and different soil types. The Collaborative Research Support Programs (CRSP) also support networks for their specific topics, for example, small ruminants, bean-cowpea, sorghum, millet, tropical soils, and other single issues. The International Agricultural Research Centers have also developed networks with National Agricultural Research Centers for different commodities and problems. These networks work because they link people with common interests and there is a framework within which they can operate. Networks that serve to draw components together to focus on more complex field problems are harder to develop but all the more needed. In fact, strong links between research organizations and users, as suggested in the preceding section, are a necessary element for successful networks (Table 4.1).

Client involvement is essential to the research priority-setting process. One example of a client-need-driven network is the Women in International Development (WID) network. The network was developed to focus more attention on gender issues in development, for instance, equal access to credit, land, and services. The network in the United States has membership dues structure, annual meetings, and publications. It also interacts with similar networks around the world and with a focal point in the United Nations. It grew from a number of initiatives in Third World and devel-

TABLE 4.1   Some Principles of Success for Networks

1. The problem is widely shared.
2. Participants are motivated by self-interest.
3. Participants are involved in planning and management of the network.
4. The problem or focus of the network is clearly defined.
5. A baseline study is undertaken to produce an authoritative founding document.
6. A realistic research agenda is drawn up.
7. Research and management are flexible.
8. The network is constantly infused with new ideas and technologies.
9. Regular workshops or conferences are held to provide opportunities for assessing progress and discussing problems.
10. Collaborators contribute resources.
11. External funding is provided to facilitate travel, training, and meetings.
12. Collaborators have sufficient training and expertise to contribute effectively.
13. The network's membership is relatively stable.
14. Leadership is efficient and enlightened.

Source: Plucknett et al., 1990

oped countries. The network was not imposed on the members—in fact, its success is based in part on a comradery formed of common interests and problems. It is potentially self-supporting but attracts funds from foundations and governments.

A network similarly structured and motivated might help address the needs for soil and water research. Although the WID network is still in its infancy, it has already identified important research needs, has had significant impacts on projects, and has developed an impressive body of literature. A similar organization could help define research priorities, provide feedback on the applicability of research findings, and integrate soil and water research into efforts to solve other resource management problems.

MECHANISMS TO EVALUATE THE EFFECTS OF RESEARCH

An often neglected element in the setting of research priorities is the evaluation of existing research efforts to get feedback about the relative success of current directions. Evaluators should try to assess whether there is evidence of progress toward sustainability as a result of the research. While sustainability—a slippery concept, at best—is difficult to measure, a range of questions could be posed that could serve as indicators. For instance, evaluators should be able to ask whether the research is actually making some difference in the field. Are people doing or trying different things as a result of the research? If they are not changing, why not? Are soil and water properties improving? Are rates of degradation slowing?

Some mechanism is needed to allow evaluators to go into the field and assess impacts and effectiveness. It should be possible to evaluate the degree of dissemination of research results and at least speculate on the potential impacts in the field.

Many research programs include external evaluations of how the activities are administered.[1] However, in addition, field-level evaluations are also needed to ascertain the actual level of dissemination of the relevant research. These field evaluations should be carried out by interdisciplinary teams that include at least one social and one physical scientist. The teams should focus on analyzing farmers' awareness of the new information being generated and disseminated as well as on the apparent effectiveness of any new practices that farmers are using as a result of research. This evaluation of field studies would also provide useful feedback to help guide the evolving research process.

MECHANISMS TO SET RESEARCH PRIORITIES

Although chapter 3 presented an agenda of research priorities as requested by AID, the committee unanimously agrees that research priorities will change, sometimes dramatically, over time, and the list only represents a first view of a longer research trajectory. Long-term benefits can accrue only if some mechanism is established to guide an ongoing evaluation and review of priorities.

Such priority-setting processes should involve a standing oversight committee that would evaluate existing field research and make judgments about the relative merits of the research efforts, in light of current needs and philosophies. Periodic workshops could be held, perhaps every three to four years, to assess progress and redefine priorities. A number of possibilities for such an oversight committee exist, such as a committee of the National Research Council, or an organization like the Board for International Food and Agriculture Development and Economic Cooperation. An alternative might be to involve the Collaborative Research Support Program, which was recently recommended (NRC, 1991b) that focuses on

---

[1]For instance, an external evaluation of a research program is called for in the National Research Council's recent report on a CRSP devoted to sustainable agriculture and natural resource management (NRC, 1991b). The external evaluation is program, rather than field, oriented: "The external evaluation panel will consist of a minimum of three senior scientists recognized by their peers and selected . . . for expertise relevant to the SANREM program and experience in research or research administration. The responsibility of the panel will be to evaluate, as deemed necessary, the status, funding, progress, plans, and prospects of the SANREM program and make recommendations based on these evaluations. . . ."

sustainable agriculture and natural resource management.[2] A committee associated with that CRSP could be assigned the task of reviewing and revising research priorities.

MECHANISMS TO INTEGRATE DATA REQUIREMENTS FOR SITE-SPECIFIC ANALYSIS

When matching crop requirements to the characteristics of the land and its owner, agriculture is part art and part science. The success of agriculture depends on the degree to which the match between crops and land is successful, and one of the roles of the farmer is to rectify any mismatches in a way that is economically feasible and environmentally sustainable. In modern agriculture, the performance of a crop planted at a particular site and time of year generally can be judged if the requirements of the crop and the characteristics of the land are known. The two greatest influences on production are soil and weather. To predict production at a particular site, therefore, a minimum set of soil, crop, and weather data is required. (It bears repeating, however, that in Third World areas where labor availability, price fluctuations, and transport problems exist, such efforts have only modest predictive powers.)

For instance, in 1982 an international group of agricultural systems scientists met at the International Crop Research Center for the Semi-Arid Tropics in Hyderabad, India, to specify the minimum data set needed to predict crop performance in any agroclimatic zone. The participants used systems analysis and crop simulation models as the primary means to match crop requirements to land characteristics. The aim was to develop a foundation for dealing with the soil-plant-atmosphere continuum so that strong links could later be forged between the biophysical and socioeconomic processes.

The scope of work was limited to 10 food crops including 4 cereals (maize, rice, sorghum, and wheat); 3 grain legumes (dry beans, groundnut, and soybean); and 3 root crops (aroid, cassava, and potato). The report

---

[2]Existing CRSPs conduct research on widely diverse issues, including: (1) fisheries stock assessment, (2) human nutrition, (3) beans and cowpeas, (4) peanuts, (5) pond dynamics and aquaculture, (6) small ruminants, (7) sorghum and millet, and (8) tropical soil management. These programs involve more than 700 experienced international scientists from 32 U.S. universities and 80 international research institutions. The design of each CRSP reflects the assumption that international collaboration is key to successful agricultural research with the host country and U.S. researchers sharing in the identification of research needs, the design of experiments, and the analysis of results. Opportunities for training and improved researcher-to-researcher links are also part of the CRSP mission.

defines the minimum data set needed to match crop requirements to land characteristics and serves as a guide for designing field experiments for model validation and model application (International Benchmark Sites Network, 1988). If a particular variety is found unsuitable to a farm, the crop genetic coefficients allow other crops and crop varieties to be examined. An economic assessment can be made for each management option as long as input costs and market prices are available. Although this minimum data set needs to be expanded to include a wider range of soil and water characteristics such as data needed to assess salinity, sodicity, metation toxicities, and nutrients such as phosphorus, calcium, magnesium, sulfur, potassium, and trace elements, it is an example of one careful attempt to improve our capability to identify appropriate crops for specific sites.

Dynamic, process-based crop-environment simulation models can be developed by interdisciplinary teams. These models are generic and are designed to operate anywhere in the world. The minimum set of soil, crop, and weather data enables the models to generate site-specific information for decision support. Models focused on human systems also offer potential, and the process of designing them could help delimit the critical human behaviors and beliefs that influence agricultural decision making and practice. The collection or compilation of the minimum data set for systems simulation should be the responsibility of the user nation. Countries unable to establish a data base for their soil and climate will require assistance from development agencies. The compilation of genetic coefficients should be the responsibility of plant breeders (Hunt et al., 1989). Although these types of efforts are certainly not panaceas for bringing about sustainability, they are useful tools in any broad effort to develop and implement better agricultural management.

CRITICAL COMMUNICATIONS ISSUES

Promoting sustainable agriculture and the wise use of soil and water resources will require more than scientific innovation. Any integrated research strategy will need to find ways to enhance communication among the different actors, from researchers to farmers. As we have learned from experience in the U.S. land-grant system, research alone cannot be expected to carry the load of agricultural development and natural resource conservation. The adoption of new policies, technologies, and farming systems will require social change—the kind of change that can be encouraged by effective communication. Certain key communication issues must be considered (Caudle, 1990):

1. Those who most urgently need information do not all speak the same language—literally or figuratively. They do not see resources the same

To be sustainable, agricultural research strategies must include dynamic interaction and communication between scientists and farmers. These Kampuchean rice farmers are taking a training course at the International Rice Research Institute in the Philippines. *Credit*: International Rice Research Institute.

way, they do not enjoy equal access to productive resources, and tenure and access rights are mediated through complex social relationships of kin, cohort, and obligation. They number in the hundreds of millions, and their cultures, levels of education, needs, and expectations are exceedingly diverse. These audiences, often illiterate, cannot always be reached through the traditional channels used to communicate with farmers in the United States because extension systems in developing countries often are inadequate. These farmers, many of whom are women, do not always toil individually. They often are members of different kinds of institutions ranging from extended families to peasant leagues that could be used as vehicles for transmitting information.

2. Information about agricultural technologies and soil and water resources primarily consists of formal research results from individual projects or experiments and vast informal practical knowledge maintained and transmitted by farmers. While the "language of science" may be universally understood in the community of researchers, the ways local knowledge is encoded and transferred is tied to local idioms and institutions. To provide a place for these formal and traditional knowledges to meet will require scientists to make the effort to understand their clients. Many well-intentioned scientists have gone about their work assuming that their clients

are basically similar to their U.S. counterparts. Cultural, economic, class, and gender differences that affect the utility of research were often overlooked, and the logic of existing practices was often misunderstood. Any strategy for sustainability must rely in the long run on a much more dynamic interaction and communication between scientists and farmers. Special mention should be made of the importance of women farmers and farming systems, so that research strategies include programs that substantially address their concerns.

There are several means for achieving these newly focused efforts, including careful analysis of local societies and economies so that how decisions are made can be understood in light of gender concerns, economic strategy, and household strategy. Another means for promoting this holistic approach is the use of computer-based decision-support systems, including simulation models and expert systems. These systems not only have the potential to store, organize, and access huge amounts of data, they also provide the opportunity and impetus for applying scientific information.

Whatever the medium of communication, we should recognize that disciplinary specialization is not always the ally of development. In some cases, this specialization has led to a confusion of tongues. When we approach the people of a foreign country and offer them our help, we must all speak the same language: theirs. Although the importance of using local languages seems obvious, it is too often overlooked when time and money are short.

3. In an era of intense competition for diminishing public support, there is a premium on "success stories"—on results that may be used to tout a program to constituents and thereby increase its funding or prolong its life. Unfortunately, the results of soil and water research typically do not lend themselves to the development of simple, easily understood "products." Such research tends to produce incremental and sometimes subtle improvements, not the dramatic breakthroughs that attract publicity. Agencies should recognize that in the world of science communication, not all subjects are created equal. While investments in medical science regularly pay off with dramatic stories and media attention, investments in soil and water science may never yield a single headline. Professionals could advise program managers and funding agencies on alternative strategies for public relations.

4. Most of the resources now expended on communications in agriculture and natural-resource programs are devoted to "in-house" publications such as newsletters and project reports, and the bulk of these are directed at the funding agency to meet documentation requirements. Typically, communications professionals are involved only in the final production of a publication, not in its planning or development. For almost every program there is a bale of brochures, reports, papers, proposals—many of them unread, ill-conceived, and ineffective in the cause of helping developing countries.

Communications dollars should be spent on products that will put information in the hands of people who need it, in a form people can use. This
means placing professional communicators on staff, and making them integral to the operation, using their input in all aspects of program development and management. It does not mean hiring an occasional consultant.
Agencies should have one staff communicator review and comment on
every program or project proposal.

5

Conclusions

This committee was charged to review the constraints on productive and environmentally sound soil and water management and to outline a research strategy for addressing these problems in the developing countries. The imperative for this activity arises from the ever-increasing pressures on soil and water resources caused by escalating world populations and changes in markets. The implications of possible changes in the global climate adds another dimension to these stresses.

The committee finds that the fundamental problems of soil and water management are predictable and, in many cases, solutions can be developed. However, soil and water management are not, and cannot be, "stand alone" issues. Rather, they are components in the overall fabric of resource use to meet basic human needs, and thus should be addressed within the context of whole farm and overall societal interests. Consequently, soil and water research should interact with all elements of the land use system and should be focused not only on solving immediate crop production problems but also on ensuring long-term preservation of these vital natural resources. Continued, increased attention to population issues, of course, is also critical.

Accomplishing this goal will require changes in our traditional approach to problem-solving. Researchers increasingly must cross the boundaries of their individual disciplines; they must broaden their perspective to see the merits of local forms of knowledge; and they must look to the farmer—the ultimate client for their information—for help in defining a practical context for research. This change in vision is under way to various degrees already, but the pace of change is slow.

This chapter presents highlights of the committee's deliberations on the mechanisms needed to speed progress toward sustainable approaches to agriculture and natural resource management. This framework of ideas is presented as broad conclusions rather than detailed recommendations with the hope that they will spark internal debate, and perhaps change, within the relevant research organizations and development assistance organizations. The issues fall into five categories: (1) the better use of existing knowledge; (2) the importance of new approaches to research that link traditionally discrete elements; (3) the use of both scientific and indigenous knowledge; (4) the need for a periodic reassessment of priorities for research; and (5) current high priority research areas.

Use of Existing Knowledge

A major constraint on progress in solving soil and water problems in the developing world is the limited application of known principles and techniques. Years of experience and experimentation have given rise to a fairly good understanding of these basic principles, yet our ability to apply this knowledge to solve problems in the complex local and cultural settings of the tropics is weak. One reason for this deficiency is the difficulty encountered in transferring information between temperate and tropical environments. Judging what information is appropriate takes a special understanding of both environments and their people. Another important factor is institutional: sustainable agriculture is not a high priority within many universities and research institutions, whether in the developing or developed world. Therefore:

• Research organizations and individual researchers should be encouraged and funded to synthesize available information into formats and languages useful to policymakers for strategic planning and forms useful to activists working at the production level (e.g., extension agents, representatives of private voluntary organizations, and producer associations and cooperatives).

The Need for a Change in the Research Paradigm

Traditional, typically single-discipline approaches to research brought great progress in the past, but this research paradigm is inappropriate for the types of problems that must be addressed to develop and implement sustainable agriculture and natural resource management. Developing sound land use practices for marginal lands, expanding production on quality lands, and restoring degraded lands all require a systems approach to research. Single-discipline research has an inherent limitation when used to address problems with complex scopes. Past approaches also tended to distance the researcher from the ultimate recipient of the information—the farmer. These problems affect the whole research establishment and not isolated funding agencies. Ways must be found to encourage more collaborative, interdisciplinary research, and to reward efforts to investigate the linkages between components that make sustainability questions particularly complex. The shift in paradigm should encourage more on-farm interactions and a vision of the farm not just as a producer of commodities but as a system. In particular, the links between physical and social issues, between agriculture and environmental issues, and between commodity approaches and ecosystem approaches deserve attention. Therefore:

• Research organizations should develop incentives to encourage collaborative, interdisciplinary research focused on bridging the gap between scientific principles and field-level application of knowledge.
• Universities, other research organizations, and professional societies should, as a matter of urgency, increase professional recognition for contributions in interdisciplinary and international areas.
• Specialists working in the commodity-focused CRSPs should seek ways to integrate their knowledge into work on sustainability questions.
• Research organizations should work to develop integrative, systems models to address the soil and water problems critical to long-term natural resource management.

Links Between Scientific and Indigenous Knowledge

When two libraries are available, each containing a different wealth of information, both should be used. Many insights about methods of managing agriculture and natural resources are available from local people. Although trial and error is a slow process, it has served farmers for generations and still has lessons to share. Often, indigenous knowledge can guide researchers in the selection of pertinent research questions. It can provide a base upon which more analytical, precise scientific investigations can be built. Indigenous knowledge and scientific knowledge can be blended to bring about practical new techniques. What is needed is more than simple inventories, it is the additional steps of analysis and evaluation that the scientific method can provide. Yet, in most cases these types of knowledge are rarely combined in effective ways. Therefore:

• Research organizations should sponsor efforts to document and scientifically validate local indigenous knowledge. The documentation should include collection and analysis of indigenous methods and technologies, such as information on classification of soils, water harvesting techniques, land management protocols, and cropping systems. Promising technologies should be evaluated and the possibilities for local improvement and adaptation to other locations explored.

The Process of Setting Priorities

Although it is important to define a clear set of priorities to guide soil and water research, the process of setting those priorities may in time prove to be more important than any specific list of issues. Priorities change over time. Agricultural systems evolve, as does our knowledge of them. Hence it is critical that the selection of research priorities be an ongoing process. Research priorities should be reviewed periodically and adjusted to serve the most urgent issues at hand. A framework is necessary to evaluate and revise soil and water research priorities to keep them flexible and responsive. The process should include mechanisms so the research is driven by client need, which would require a carefully designed feedback system. Therefore:

• A clear mechanism should be established to ensure that a recurrent priority-setting process occurs, perhaps by delegation to an existing institution that is close to the research environment.

• To be most effective, the priority-setting process should begin with attention to regional needs by defining regional priorities.

• Whatever the mechanism used for this task, it should involve the combined assessment of researchers, change agents, research users, and funding agencies.

Current Research Priorities

Increasing stress on the world's agricultural systems necessitates a strong emphasis on improving agricultural production and also makes attention to the long-term stewardship of the earth's natural resource base imperative. Sustainable agricultural systems must maintain and enhance both biological and economic productivity. They must be stable and resilient. To be acceptable to the people who must use them, such systems must be economically viable over the short and long term, and must be socially compatible with the needs of the local population as well as broader political agendas. Accomplishing these goals is no small task, and research of numerous types will be essential.

Although the realm of potentially valuable research topics is vast, some focus is necessary to speed progress. From its deliberations, this committee selects the following broad research areas for priority attention during the 1990s:

• Institutional constraints on resource conservation;
• Soil biological processes;
• Management of soil characteristics;

- Water resource management;
- Matching crops to environments; and
- Incorporating social and cultural dimensions into research.

Research to encourage the wise management of soil and water resources is critical to maintaining and enhancing agricultural productivity over the long term. The United States is in a position to provide leadership, offer technical and financial assistance, and encourage international cooperation to support this emphasis on sustainable agriculture and natural resource management. It will not be easy to implement integrated systems-based collaborative research into the ecological and socioeconomic characteristics of sustainable agriculture and natural resource management, but it is imperative.

References

Altieri, M., and S. Hecht. 1990. Agroecology and Small Farm Development. Boca Raton, FL: CRC Press.

Caudle, Neil. 1990. Communication to the Committee on International Soil and Water Research. Workshop, October 1–2, 1990.

Colfer, C. J. P. 1987. Soil Science and Soil Management: An Anthropologist's Role in the Tropsoils Project. In Farming Systems Support Project Newsletter (FSSP), Volume 5, Number 1.

Cuc, L. T., K. Gillogly, and A. T. Rambo. 1990. Agroecosystems of the Midlands of Northern Vietnam. A Report on a Preliminary Human Ecology Field Study of Three Districts in Vinh Phu Province. Environment and Policy Institute, East-West Center Occasional Paper No. 12.

Department of State. February 1990. Outgoing telegram, "Information message on sustainable agricultural development."

Doyle, D. 1991. Sustainable Development: Growth Without Losing Ground. Journal of Soil and Water Conservation. January–February 1991.

Edwards, C. A., R. Lal, P. Madden, R. H. Miller, and G. House, eds. 1990. Sustainable Agricultural Systems. Ankeny, Iowa: Soil Conservation Society of America.

Grove, T. L., C. A. Edwards, R. R. Harwood, and C. J. Pierce Colfer. 1990. The Role of Agroecology and Integrated Farming Systems in Agricultural Sustainability. Paper prepared for the Forum on Sustainable Agriculture and Natural Resource Management, November 13–16, 1990, National Research Council, Washington, D.C.

Hunt, L. A., J. W. Jones, J. T. Ritchie, and P. S. Teng. 1989. Genetic coefficients for the IBSNAT crop models. In the Proceedings of the International Benchmark Sites Network for Agrotechnology Transfers (IBSNAT) Symposium. 81st Annual Meeting of the American Society of Agronomy, Las Vegas, Nevada. Department of Agronomy and Soil Science, University of Hawaii, Honolulu, HI.

International Benchmark Sites Network for Agrotechnology Transfer. 1988. Experimental Design and Data Collection Procedure for IBSNAT: Minimum data set for systems analysis and crop simulation. 3rd edition. Department of Agronomy and Soil Science, University of Hawaii, Honolulu, HI.

Jodha, N. S. 1990. Sustainable mountain agriculture: Some predictions. Paper prepared for the Forum on Sustainable Agriculture and Natural Resource Management, November 13–16, 1990, National Research Council, Washington, D.C.

KEPAS. 1985. The Critical Uplands of Eastern Java: An Agroecosystems Analysis. Kelompok Penelitian Agro-Ekosistem, Agency for Agricultural Research and Development, Republic of Indonesia. Report prepared by Gordon R. Conway, Ibrahim Manwan, David S. McCauley, Frederich C. Roche, M. Iksan Semaoen, and M. Winarno.

Lal, R. 1988. Soil degradation and the future of agriculture in sub-Saharan Africa. Journal of Soil and Water Conservation 43(6):444–451.

Marten, G. G., and A. T. Rambo. 1988. Guidelines for writing comparative case studies of Southeast Asian rural communities. In Agroecosystem Research for Rural Development. Appendix 3. Edited by K. Rerkasem and A. T. Rambo. Chiang Mai, Thailand: Multiple Cropping Centre and SUAN.

National Research Council. 1974. More Water for Arid Lands: Promising Technologies and Research Opportunities, Washington, D.C.: National Academy Press.

National Research Council. 1981a. The Water Buffalo: New Prospects for an Underutilized Animal, Washington, D.C.: National Academy Press.

National Research Council. 1981b. The Winged Bean: A High-Protein Crop for the Tropics, Washington, D.C.: National Academy Press.

National Research Council. 1983a. Amaranth: Modern Prospects for an Ancient Crop, Washington, D.C.: National Academy Press.

National Research Council. 1983b. Calliandra: A Versatile Small Tree for the Humid Tropics, Washington, D.C.: National Academy Press.

National Research Council. 1983c. Casuarinas: Nitrogen-Fixing Trees for Adverse Sites, Washington, D.C.: National Academy Press.

National Research Council. 1983d. Little-Known Asian Animals with a Promising Economic Future, Washington, D.C.: National Academy Press.

National Research Council. 1984. Leucaena: Promising Forage and Tree Crop for the Tropics, 2d ed. Washington, D.C.: National Academy Press.

National Research Council. 1985. Jojoba: New Crop for Arid Lands, Washington, D.C.: National Academy Press.

National Research Council. 1989a. Alternative Agriculture, Washington, D.C.: National Academy Press.

National Research Council. 1989b. Investing in Research: A Proposal to Strengthen the Agricultural, Food, and Environmental System. Washington, D.C.: National Academy Press.

National Research Council. 1989c. Saline Agriculture: Salt-Tolerant Plants for Developing Countries, Washington, D.C.: National Academy Press.

National Research Council. 1991a. Microlivestock: Little-Known Small Animals with a Promising Economic Future. Washington, D.C.: National Academy Press.

National Research Council. 1991b. Toward Sustainability: A Plan for Collaboration Research on Agriculture and National Resource Management. Washington, D.C.: National Academy Press.

Office of Technology Assessment. 1988. Enhancing Agriculture in Africa: A Role for U.S. Development Assistance. OTA-F-357. Washington, DC: US Government Printing Office.

Pimentel, D., J. Allen, A. Beers, L. Guinand, R. Linder, P. McLaughlin, B. Meer, D. Musonda, D. Perdue, S. Poisson, S. Siebert, K. Stoner, R. Salaziar, and A. Hawkins. 1987. World agriculture and soil erosion. Bioscience 37:277–283.

Plucknett, D. L., N. J. H. Smith, and S. Ozgediz. 1990. Networking in International Agricultural Research. Cornell University Press.

Sanchez, P., and J. Benites. 1987. Low input cropping for acid soils of the humid tropics. Science 238(December):1521–1527.

Toledo, J., and E. Serrão. 1982. Pasture and animal production in amazonia. Pp. 281–309 in Proceedings of International Conference, Amazonia: Agriculture and Land Use Research. Cali, Columbia: Centro Internacional de Agricultura Tropical.

APPENDIX A

Biographical Sketches of Committee Members

LEONARD BERRY, *Chair*, received his B.Sc., M.Sc., and Ph.D. from the University of Bristol, England. He is currently Provost and Academic Vice President of Florida Atlantic University. Dr. Berry is a member of many professional societies including the Royal Geographical Society, the Association of American Geographers, and the Institute of British Geographers. He has also served on numerous committees for the National Research Council.

SUSANNA B. HECHT is currently an associate professor in the Graduate School of Architecture and Urban Planning at the University of California, Los Angeles. She received her B.Sc. from The University of Chicago, and M.A. and Ph.D. in geography from the University of California, Berkeley. Her areas of expertise include regional development policy and planning, resource science, soil science, and geography. She is a specialist in the development of the humid tropics. Dr. Hecht is a member of the American Anthropological Association, the Society of Tropical Foresters, and the Society of Conservation Biology.

CHARLES W. HOWE received his B.A. from Rice University, and his M.A. and Ph.D. in economics from Stanford University. He is currently professor of economics at the University of Colorado. Dr. Howe is also the President of the Association of Environment and Resource Economists.

JACK KELLER received his B.S. from the University of Colorado, M.S. from Colorado State University, and his Ph.D. in irrigation engineering from Utah State University. He is currently teaching at Utah State University and performs educational activities for governments, schools, and industry in on-farm design irrigation throughout the United States and a number of other countries. Dr. Keller is also Co-Director of the Water

Management Synthesis Project under the U.S. Agency for International Development. He is a member of the National Academy of Engineering and a number of professional organizations.

CHARLES B. McCANTS received his B.S. and M.S. from North Carolina State University, and his Ph.D. in soil science from Iowa State University. He is professor emeritus at North Carolina State University where he served as Director of the Management Entity for the Soil Management Collaborative Research Support Program, Head of the Soil Science Department, and research leader in soil management.

HUGH POPENOE received his B.S. from the University of California, Davis, and his Ph.D. in soils from the University of Florida. He is currently Director of International Programs in Agriculture; Director of the Center for Tropical Agriculture; professor at the University of Florida; and President of the American Water Buffalo Association. His area of expertise is on tropical land management and tropical ecology. Dr. Popenoe has served on numerous National Research Council committees.

GORO UEHARA is currently a professor in the Department of Agronomy and Soil Science at the University of Hawaii. He received his B.S. and M.S. from the University of Hawaii, and a Ph.D. in soil science from Michigan State University. His research intent is to develop simulation models to predict crop yields in any location in the world. Dr. Uehara is a member of the Soil Science Society of America, the Crop Science Society of America, the American Society of Agronomy, and the American Association for the Advancement of Science.

APPENDIX B

Workshop Participants
October 1–2, 1990
Irvine, CA

The committee's members wish to extend their sincere appreciation to all the people who shared their expertise and experience with us during the preparation of this report. Our special thanks go to the people who participated in the October 1990 information gathering workshop. The workshop gave the committee access to a broad range of ideas, and set the stage for the themes developed in this document. In the end, however, the committee claims sole responsibility for the content and recommendations of this report.

Peter Ahn
Director, African Vertisols Project
International Bureau for Soils
    Research and Management
P.O. Box 23001
Nairobi, Kenya

Miquel Altieri
Biological Control
University of California
Berkeley, California 94704

Ben B. Bohlool
NifTAL Project
1000 Holomua Road
P.O. Box O
Paia, Maui, Hawaii 96779-9744

Elmer Bornemisza
Apartado 1166-1000
San Jose, Costa Rica

Judy Carney
Department of Geography
University of California
Los Angeles, California 90024

Neil Caudle
Department of Agricultural
    Communications
North Carolina State University
Raleigh, North Carolina
27695-7603

Carol J. Pierce Colfer
512 SW Maplecrest Drive
Portland, Oregon 97219

Pierre Crosson
Resources for the Future
1616 P St., N.W.
Washington, D.C. 20035

John Duxberry
Department of Soil, Crop, and
   Atmospheric Sciences
Room 917, Bradfield Hall
Cornell University
Ithaca, New York 14853

Samir El-Swaify
Department of Agronomy and Soil
   Science
1910 East-West Road
University of Hawaii
Honolulu, Hawaii 96822

Dale Harpstead
Crop and Soil Sciences
Michigan State University
East Lansing, Michigan 48824

Donald Humpal
Development Alternatives, Inc.
4811 Chippendale Drive, Suite 702
Sacramento, California 95841

Tony Juo
Department of Soil and Crop
   Sciences
Texas A&M University
College Station, Texas 77843-2474

Edward Kanemasu
Agronomy Department
University of Georgia
College of Agriculture
Georgia Station
Griffin, Georgia 30223-1797

Jon R. Morris
Sociology and Anthropology
   Department
Utah State University
Logan, Utah 84322-4105

Don Nielson
University of California, Davis
Department of Agronomy and
   Range Sciences
Davis, California 95616

Cheryl Palm
Soil Science Department
North Carolina State University
Raleigh, North Carolina
27695-7619

Jot Smyth
Soil Science Department
North Carolina State University
Raleigh, North Carolina
27695-7619

Philip Thornton
Agro-Economic Division
IFDC
P.O. Box 2040
Muscle Shoals, Alabama 35662

Thomas Weaver
Department of Resource
   Economics
Lippitt Hall
University of Rhode Island
Kingston, Rhode Island 02881

Charles Wendt
Texas Agricultural Experiment
   Station
Route 3
Lubbock, Texas 79401

Lyman S. Willardson
Agricultural and Irrigation
   Engineering Department
Utah State University
Logan, Utah 84322-4105

Russell S. Yost
Department of Agronomy
  and Soil Science
1910 East-West Road
University of Hawaii
Honolulu, Hawaii 96822

AID Representatives

David Bathrick
James Bonner
William Furtick
Thurman Grove
Raymond Meyer

NRC Representatives

Jeanne Aquilino
Michael McD. Dow
Chris Elfring
John Hurley
Stephen D. Parker